How to Write and Present Technical Information

How to Write and Present Technical Information

Fourth Edition

Charles H. Sides

An Imprint of ABC-CLIO, LLC
Santa Barbara, California • Denver, Colorado

Library of Congress Cataloging-in-Publication Data

Names: Sides, Charles H., 1952- author.
Title: How to write and present technical information / Charles H. Sides.
Description: Fourth edition. | Santa Barbara, California : Greenwood, [2017] |
Includes bibliographical references and index. |
Identifiers: LCCN 2017019332 (print) | LCCN 2017022948 (ebook) |
ISBN 9781440855061 (ebook) | ISBN 9781440855054 (alk. paper) |
ISBN 9781440855078 (pbk. : alk. paper)
Subjects: LCSH: Technical writing. | Communication of technical information.
Classification: LCC T11 (ebook) | LCC T11 .S528 2017 (print) | DDC 808.06/66—dc23
LC record available at https://lccn.loc.gov/2017019332

ISBN: 978-1-4408-5505-4
EISBN: 978-1-4408-5506-1
Paperback ISBN: 918-1-4408-5507-8

21 20 19 18 17 1 2 3 4 5

This book is also available as an eBook.

Greenwood
An Imprint of ABC-CLIO, LLC

ABC-CLIO, LLC
130 Cremona Drive, P.O. Box 1911
Santa Barbara, California 93116-1911
www.abc-clio.com

This book is printed on acid-free paper ♾
Manufactured in the United States of America

Contents

PART IV: HOW TO WRITE SPECIFIC DOCUMENTS

PART V: HOW TO WRITE AND DESIGN FOR DIGITAL MEDIA

PART VI: HOW TO WRITE AND DESIGN ASSOCIATED COMMUNICATIONS

PART VII: FINISHING YOUR WORK

PART VIII: PRESENTATIONS AND MEETINGS

PART IX: CONCLUSION

Preface

Nothing better indicates the need for a fourth edition of *How to Write and Present Technical Information* than this: when the third edition appeared, the World Wide Web was three years old. In each edition, dating back to the first one in 1983, I have pointed out that technical writing is not reality but a re-creation of reality to meet the needs of readers. Even with the vast changes that have occurred in technological industries around the world in recent years and even with the greater changes predicted as the 21st century matures, the theme of this fourth edition remains much the same. Professionals in technological disciplines simply have no time for minutiae in a report or paper. They have even less time for poorly written rambling accounts of technical information. Between 24/7 global interconnectedness that is instantaneous, between always-on cell phones and videoconferencing, today's professionals more than ever are hard-pressed to stay up with the real-time information flow of today's technological world. For example, a start-up company in the restaurant service industry is founded on a product that provides up-to-the-second analyses of their clients' services, so that adjustments in inventory, service, etc., can be made immediately, at any time throughout the world. But even in harried communication environments such as these, professionals still need to communicate, perhaps even moreso. They need to know as quickly and as easily as possible, what the main points of communication are, why these points are important, and what they—the readers—should do with them. So despite the fact that global interaction in professional disciplines is occurring at never-before-accomplished speeds, and despite the fact that the means of communication change quickly with every newly introduced social media platform; the standards of successful communication are comparatively permanent.

As a result, even more so today, we recognize that not all of us read, write, and work equally well in the same fashion. We each have a style that suits us, and one of the goals of this book remains to help individuals understand their work styles, the work styles of colleagues and customers, and the roles these styles play in written communication. For example, some people are "note writers." Tablets and smart phones are crucial in this approach to professional organization. Many people write down every scrap of information they encounter when researching a project. Everything is neat, organized, and easily accessible. Other people (myself included) are "head writers." These people do not write down much of the information they research but mull it over in their heads, organizing and reorganizing (often for weeks) before they put word to paper. It's pretty easy to tell when someone whose style resembles the first example is working. There is tangible evidence of it. But it's much more difficult to tell in the second example. These people appear to be daydreaming or goofing off, lost in whatever reverie they are inhabiting.

Remember, however, both styles work! And if you are comfortable with one or the other, don't let people talk you out of it simply because it is not their style. In other words, watch out for people who have the "do it right, do it my way" syndrome. They're dangerous, and they might cause you to do less than your best work. For example, for each edition of this book, I was given between six and nine months to produce a draft. In each case I spent all but the last six weeks thinking about the subject, planning it, and organizing it—in my head. Once again, it's now approximately six weeks before the draft is due. Will I finish? You bet! Will I be pleased with what I have done? Absolutely! This style works for me, and it may work for you. If so, use it. If not, don't. We will look at the issue of work styles more in Chapter 1 of this book.

A third theme incorporated into the fourth edition of *How to Write and Present Technical Information* specifically addresses an ongoing economic reality in 21st-century technological industries—downsizing. Consequently, a key principle of this edition is that the person who brings the most skills to the position is the person who remains in that position, while others do not. Writing about and presenting technical information today involve skills that were rarely considered part of the technical communicator's tool kit several decades ago when I began my career. These tasks include writing public relations materials, marketing and advertising materials, web content, social media posts, and designing multimedia communications. Each of these duties, as well as mastering the software platforms that support them, has become an integral part of the job of a person who writes and presents technical communication.

So, the fourth edition of this book continues to be intended for people who work within technological disciplines, but in the 21st century, all disciplines are technological disciplines. What the book now contains can be used equally well in virtually any discipline. Basically, it offers common-sense advice on how to write professional communications that do not fail. The book could be used in a college technical writing course; the first three editions were so used regularly. It could be used in writing across the curriculum or writing within the discipline courses. But none of the previous editions, nor this one, is specially aimed at that academic audience, since these books lack things such as exercises, case studies, discussion topics, etc. The fourth edition, as was the case with each previous edition, is primarily for you—professionals who have to write documents as quickly and as effectively as you can within whatever discipline you find yourselves.

Decades ago, the first edition of this book, *How to Write Papers and Reports about Computer Technology*, was referred to as "the most product-oriented book on technical writing on the market." I took that as a compliment—even though the statement was, I'm sure, not intended to be. Technical writing pedagogy has at times borrowed too heavily from the process approaches to teaching freshman composition. These approaches are excellent educational tools for students who are just entering college, many of whom have less than rudimentary writing skills. But the process approach to writing is best taught in the primary grades, thereby preventing large categories of writing problems before students arrive at a college. The approach does not work well for people who have to write a lot with very little time in which to do it—in other words, for professionals like you. On the job, most of us do not have the time to write languorously until we discover our purpose for writing regardless of how valuable an educational exercise that might be. The result in our professional environments would be that almost everything we wrote before our purpose came to us would have to be discarded as useless.

So, yes, writing within professional environments means viewing writing as producing a product, which many disciplines actually refer to as "the information product." That phrase makes sense. In this fourth edition of *How to Write and Present Technical Information*, you will find out how to design, organize, and write an ever-widening variety of information products.

As a result, think of this book as a reference tool. While readers might read it cover-to-cover, it is designed for you to locate specific sections that provide advice for meeting your immediate communication needs. As professionals, we will continue to find that our communication needs include designing and writing new types of messages that we never previously considered for professional environments—particularly how to use social media for professional communications. If we have learned anything from the past, it is that the future will expect more from us. This book will help you accomplish it.

Introduction

SCENARIO 1

A short while ago, you were hired by an internationally known agricultural products firm that is planning to enter the highly competitive robotic market with a revolutionary system they are calling a farmbot. As a software engineer fresh out of college, you were hired to work in the Software Systems Laboratory, designing and coding the new AI system and embedding it into an autonomous learning scheme that will be transparent to users, while concurrently permitting the bot to adapt to environmental conditions as it encounters them. Initially thrilled at such a revolutionary opportunity and challenge that has the potential to change agriculture world-wide, you have discovered that as much as 40 percent of your time is spent communicating to others what you are doing—writing specifications and weekly status reports, making presentations in company meetings, interviewing people in marketing, and talking on the phone. Not only is this work not what you were led to expect while you were in one of the country's leading technological institutes, but also you realize that freshman English—which you struggled through and frankly hated—did nothing to prepare you for the rigors of on-the-job communication.

SCENARIO 2

You have worked in the high-tech industry for several decades, rising to the upper levels of management in the corporate communication division. You've seen a lot of changes, including the dot-com boom and bust, as well as the Great Recession. A lot of people have come and gone; there have been tremendous advances in the field, and periods of economic uncertainty. Your company has survived the bust period of a few years back—just barely; and a new CEO with a reputation for off-shoring key company components has been hired to implement across-the-board downsizing and re-structuring to get things back on track. As usual in management changes, she is shaking things up, putting her own stamp on company operations, and (you fear) completely upending company culture before she heads off to her next opportunity, a rumored U.S. cabinet position. But for you there's a difference: She's told you that your division's information products are unsatisfactory and that they will be improved. You realize that riding out your career the next few years to retirement will increasingly depend on your ability to satisfy whatever she means by clear, usable communication.

SCENARIO 3

After months of hard work, long hours, and sleepless nights, you have made a key contribution to perfecting your company's advance in mobile gaming devices. The product, which has been beta tested with customer bases from preteen to adults, will be shipped next month, and the advertising people are already touting it in the trade magazines as the next Angry Birds®. The division manager has selected you, naturally, as the best person to write a public relations article for one of these magazines. Of course, you're honored. But you don't know the first thing about public relations or about writing a journal article, never having thought of yourself as much of a writer.

SCENARIO 4

Your company has decided to send you to the industry's big, annual, international conference in Geneva next month. You will be expected to make a formal presentation before an audience of at least 150 top industry professionals from around the world. Even giving short presentations at small, informal group discussions makes you nervous. The thought of standing up in front of a large crowd in an auditorium is terrifying.

Each of these scenarios is common to professionals in high-tech industries. Many people, however, find themselves unprepared to meet the challenges of these communication tasks. They spend far more time than is necessary preparing for (and fretting about) these duties, and they get far poorer results than all this time should warrant.

If you have ever found yourself in one of these scenarios, then this book is intended for you. Even if you don't recognize any of these situations now, you will eventually encounter these communication tasks if you intend to survive, prosper, and advance in high-tech industries. This book is for you, too.

PART I

Writer, Audience, and Documentation

CHAPTER 1

Who We Are and What We Do

Who was Carl G. Jung, and what does his theory of personality types have to do with communication strategies in professional environments? The first question can be easily answered: Jung was one of the most influential psychologists of the 20th century, an influential thinker and prolific writer who, with his theory of personalities, explored psychological wellness, a largely ignored area of psychology at the time. The answer to the second question is summed up by this chapter's title, "Who We Are and What We Do," and it will be answered over the next few pages. In short, who we are greatly affects what we do, how we do it, and why.

"DIFFERENT STROKES . . ."

Jung's theory of personality types suggests that individuals' personalities are best described by how two attitudes (extraversion and introversion) and four functions (sensation, intuition, feeling, and thinking) interrelate. The psychological jargon is unimportant here; what is important is how these things work.

"Extraversion" (Jung's spelling) describes individuals who prefer to focus their lives outward to the experiences of other people and things. They tend to be active and energetic—doers. "Introversion" describes individuals who prefer to focus their lives inward to the experiences of thought and reflection. They are no less active than extraverts, but their activity is internalized and hidden from the view of others. Realize that society has associated these terms with some negative inferences. "Extraversion" does not refer to blabbermouthed boors any more than "introversion" refers to hermits. In fact, Jung originally coined these words to mean exactly what their roots suggest: "extravert" (turning out) and "introvert" (turning in).

Sensing and intuition are perceptive functions; here again, Jung uses psychological jargon to describe how people take in (perceive) information. Those who prefer sensing would rather rely on their five senses alone. Those who prefer intuition use their five senses only to gain enough information to make an educated guess about what they have perceived.

Both sensing and intuition provide the information with which we make judgments. And we do that either through feeling or thinking. People who prefer "feeling judgment" generally fit their perceptions into some preconceived value system (cultural, personal, corporate, familial, religious) before deciding what to do with the information. People who prefer "thinking judgment" usually decide upon information based solely on its merit, irrespective of fitting into a value system. In this case, too, society has colored the issue with negative connotations. "Feeling" does not mean "weak and emotional," and "thinking" does not mean "dispassionate and logical." In fact, Jung thought of both functions as rational, trustworthy, and valuable ways of making judgments.

Although this explanation is grossly oversimplified, it is about as far as Jung explored the issue explicitly in his 1923 book *Psychological Types.* He did, however, imply two other attitudes—judgment and perception—which Katherine Briggs and her daughter Isabel Briggs Myers described fully in their later work, *Gifts Differing,* that led to a psychological evaluation instrument (the Myers-Briggs Type Indicator, or "MBTI" for short). Essentially, "judgment" refers to people who would prefer bringing matters to closure. "Perception" refers to people who would prefer to keep matters open-ended. The MBTI enables people, with the aid of trained professionals, to determine their type preferences, bettering their understanding of themselves and of others.

So, again, what does this have to do with us? Simple: Four two-part preferences yield 16 personality types. The descriptions of these types can readily be found on the web, as well as short, self-directed questionnaires to determine your own type.

Understand—and this point is extremely important—that these personality types are not stereotypes into which you can be pigeonholed. In fact, Jung contended that each one of us has some aspect of every single category. It's just that we prefer to use some categories over others. A helpful comparison is handedness. Most of us are either right- or left-handed, but all of us can use the unpreferred hand. Personality type is the same. Some of us are introverted, but we are able to interact with other people on a social and professional basis. Some of us prefer thinking judgment, but we are able to use and understand value systems.

PERSONALITY TYPE AND WORK

In technology, teams rather than individuals often write the content; each writer brings different work habits and preferences to their work. Understanding those differences enables you to be a more productive team member, appreciating why you and others around you prefer to do what you do. Sensing-judging (SJ) team members may have a preference for detailed descriptions, for meeting agendas, and for carefully conceived plans. Intuitive-perceptive (NP) team members might feel more comfortable in a "seat-of-the-pants" mode. Thinking (T) team members desire rational reasons for what they are doing; feeling (F) team members like a harmonious working environment. It's common sense that you should work better doing what you prefer in a way that you prefer it. That sounds ridiculously simple, but since most people think "my way is the right way," it's not as easy as it may seem.

Videoconferencing, Internet communications, and other 21st-century electronic substitutes for face-to-face, elbow-to-elbow working environments serve only to exacerbate the challenges of personality differences and personality-based work preferences. For example, the way persons use social media in professional environments might be affected by typological preferences. An NP *might* be more likely to dash off a quick tweet or e-mail; whereas an SJ *might* carefully plan the message and review it before sending it. Both types should, in fact, review and edit prior to sending, but this is one of the important ways that typed messages can function in professional environments—those who are more prone to quick work will have to develop skills in thoroughness that often come naturally to other type preferences.

A good deal of trust is required to allow your colleagues and employees to work in a way that is productive for them but counterproductive for you. If you can do it, and particularly if management in your organization supports it, the results will astound you.

CONCLUSION

In this chapter, we have briefly considered the effects that individuals' personalities can have on their work. This material forms an underlying current for much of the rest of this book, from audience analysis to report design. In short, remember that we do not all think the same way, behave the same way, value the same things, or work the same way. Realize that this also means we do not all read the same way, either. Some read every word meticulously; others skim. One of the goals of this book is to show you ways to design reports so that they succeed for both reading strategies. It's a tall order, but we need to make the information we communicate in reports and papers accessible to as many different types of people as possible.

SUGGESTED READINGS

Jung, Carl G. *Psychological Types.* Princeton, NJ: Princeton University Press, 1971.

Keirsey, David, and Marilyn Bates. *Please Understand Me: Character and Temperament Types.* Del Mar, CA: Prometheus Nemesis Books, 1978.

Lawrence, Gordon. *People Types and Tiger Stripes.* Gainesville, FL: Center for the Application of Personality Types, 1983.

Myers, Isabel Briggs. *Gifts Differing.* Palo Alto, CA: Consulting Psychologists Press, 1980.

Sides, Charles H. "What Does Jung Have to Do with Technical Writing?" *Technical Communication* 36(2) (1989): 119–126.

www.aptinternational.org

www.capt.org

CHAPTER 2

How to Define High-Quality Documentation

Before defining high-quality documentation, it might be better to ask, what is documentation? That this basic question is necessary and the numerous answers it receives show the problems that exist in writing about technology. Is "documentation" the manuals alone? The online help screens? Animated multimedia complete with audio and motion video in an online tutorial? An avatar-based query-and-response system? The e-mails that communicate information on a daily basis and lead to reports and manuals? The reports themselves, including both hard copy and electronic versions? How about papers and articles that are published in trade magazines? Public relations and marketing materials? Small group presentations? Formal presentations? Social media communications? All of the above?

The last answer is the correct one. "Documentation," when accurately defined, embraces all acts of writing, illustrating, and speaking about high technology to all possible audiences. It is an enormous communication task, fraught with proverbial snares and minefields. This complexity is perhaps the reason why many people identify poor technical communication solely with high-tech industries. Too often, they are right.

CHARACTERISTICS OF GOOD DOCUMENTATION

Now that we have determined what documentation is, we can move on to the more important problem of identifying the characteristics of good documentation. They do exist.

Meeting the Audience's Needs

Good documentation satisfies the needs of an identified audience. This concept is the most important aspect of writing in the high-tech industries. (The next chapter is devoted entirely to the subject of finding out who your audience is and what they want to read, as well as what they need to read.) Readers come to a document—be it memo, report, manual, online help, press release, or technical paper—with certain needs. They might be mildly interested in a new development in the field; they might need to get a "go-ahead" before proceeding on a new project; they might want to see whether a project can be completed economically or whether it has a place in the projected market; or they might want to learn a programming procedure in as little time as possible. They might be interested in what is trending on social media about a company and its products or services. Whatever the readers' needs, they must be met by the communication. If they are not met, the importance of the subject, and the brilliance of the author, will be irrelevant. The communication will fail. Remember, technological advancement is useless without the ability to communicate it to customers, through an increasingly wide range of media and platforms.

Good Organization

Once the author has considered the audience, organization is the second most important concern. Good documentation is rigorously organized. In fact, organization is one of the ways audience needs can be met. But as we have seen already, different types of audiences might prefer different organizational strategies. Providing multiple pathways to information is a way to ensure that documents are useful to a broad range of readers. This organization strategy is particularly useful for designing online information and using social media platforms in professional environments.

Essentially, "documentation" is writing to be used. One rarely imagines somebody curling up in front of a cozy fire on a cold winter evening with a user manual. Technical documentation does not, therefore, have to pique the interest of readers, because readers already come to the documentation with specific needs. Readers *are* interested; otherwise they would not have picked up the document or gone to the website or clicked on the link in the first place. This built-in interest does not mean, however, that documentation writers need not be concerned with readers' interests. Everything in a document must be organized with the reader in mind. The document must present the subject in such a way that all readers can get from here to there, from being interested in the topic to knowing enough about the topic to do their jobs. This is a particular challenge to using social media professionally, especially platforms such as Twitter® that limit communications to extremely truncated numbers of characters. And readers had better be able to gain this understanding without false steps, circumlocutions, and journeys down tangential paths. One of the best ways to ensure that readers will put your document down without finishing it is to confuse them. Multimedia, web designers, and game designers even have a term for it—"seat time"—the length of time a product stimulates a user to stay seated and actively engaged.

Now, I just said that a document does not have to stir interest, but that does not mean it has to be boring. Technology is fascinating, and the only thing that could possibly be boring about it is how writers treat it in documents. Nor does this mean that documents cannot be persuasive. Subtle persuasion is a handy tool for high-tech professionals, and it can be achieved through the clear explanation of a topic in which readers are already interested. In a way, this assumed interest makes people who write about high-tech lucky. They don't have to work as hard to engage their readers. But certain stylistic techniques can be used to further enhance the topic, including a varied sentence structure, variable sentence lengths, a personal style, and humor.

Humor

Humor? What about humor? Does it have a place in technical communication? Not using humor in writing about technology is similar to the schoolmarm's admonition against starting sentences with "and," and every knowledgeable eighth-grader realizes that J. K. Rowling does it all the time.

The question is moot. Good documentation (particularly certain types of manuals and magazine articles, not so much reports and memos) is rife with subtle humor—much of it intentional. As a result, documentation writers must be the funniest people in high-tech today. Seriously, their humor, when done right, never gets in the way of the subject. And for certain audiences, it makes a difficult subject easier to understand. But remember: e-mail and social media posts, because of their characteristic shorter lengths, are especially tricky when it comes to using humor successfully. Be careful, and make sure that if you are intending to be humorous that your audience gets it. Otherwise, you might find yourself spending considerable time undoing unnecessary problems that you created with humor that wasn't.

Jargon

Good documentation uses jargon. That's right. Good documentation uses jargon. But that's heresy, isn't it—a sellout to those technocrats and bureaucrats who would obfuscate our language? Taken by itself, the statement about jargon is not entirely accurate, but it's not heresy, either. Most high-tech industries are extremely complex. Part of the reason for this complexity is that, compared to medicine, astronomy, and even quantum mechanics, the high-tech field is relatively new. A large part of the documentation produced in this field is written for other people in the industry, readers who have considerable knowledge of the subject.

This situation introduces the issue of audience, and that's the proper context in which to view the use of jargon. Will the readers share the writer's knowledge of terminology? If so, altering that terminology makes no sense. At worst, doing so oversimplifies the topic, damages its integrity, and limits the imagination of the writer. But—and it's an important "but"—authors must have a clear idea of who the audience is. Even managers in high-tech industries may not share the technical vocabulary of their engineers. In fact, some engineers do not always share the same vocabulary with other engineers; consequently, the use of jargon is audience-dependent. Remember: Always use what the audience will understand.

To paraphrase Samuel Taylor Coleridge, whatever can be written in fewer and simpler words without damaging the audience's understanding of the information is poorly written. You won't find a better guideline for the use of jargon than that.

Readability

"Readability" (or "usability") is the final characteristic of good documentation. Readability is the skeleton of all writing that works. We can't see it, but it holds the writing up and allows it to do things. Readability is achieved in a number of ways. Sentence and word length have something to do with it. So does sentence structure. Most readability formulas, such as the older but still worthwhile Gunning's Fog Index and the Flesch Reading Ease Test, are based on these criteria. (See Chapter 24, "How to Avoid Common Writing Problems," for more information on readability tests.) But these tests oversimplify the issue. Correctly structured, with a feel for rhythm and the ebb-and-flow of a phrase, even the long sentence is readable. When it is not overused, such a sentence can be a way to attain structural variety, or to emphasize weighty importance. Rather than relying on artificial means for determining readability, realize this truth: The more often you make a reader reread to understand what you have said, the less readable (and usable) your document is.

CONCLUSION

In this chapter, I have identified several basic characteristics of good documentation. Notice that I did not mention formats, except as examples in passing. Good documentation is independent of the format that contains it. At its most basic level, good documentation is good writing and good speaking. It is also good sense.

To that end, the rest of this book will examine in detail the techniques of good writing and speaking as they can be applied to informational products within the high-tech industries.

SUGGESTED READINGS

Lannon, John M., and Laura J. Gurak. *Technical Communication*, 13th ed. New York: Longman Publishing Group, 2013.

Markel, Mike. *Technical Communication*, 11th ed. New York: Bedford/St. Martin's, 2014.

Tebeaux, Elizabeth, and Sam Dragga. *Essentials of Technical Communication*, 3rd ed. Oxford: Oxford University Press, 2015.

CHAPTER 3

How to Define Your Audience

The first question that document writers might want to ask is, why is audience definition so important? The answer is simple: Audience definition helps writers target the major group of readers for any document, regardless of the medium that delivers it. Audience analysis enables writers to discover what their readers know about the subject. It focuses on what the readers need to know to perform their jobs better or to increase their knowledge about the subject. And it helps writers determine what their readers will do with the information they read. The lack of conscious audience definition is responsible for the regrettable state of much technical writing: documents with poorly defined purposes, written for readers whom the writer has not considered and does not understand.

Viewed in the proper context, audience definition is the single most important aspect of rendering technical information usable to readers. This concept is the underlying principle of this book. The system presented in this chapter will provide writers with an easy-to-use method for defining the audience.

THE AUDIENCE DEFINITION SYSTEM

My systematic method for audience definition is divided into four processes: (1) defining who the readers are; (2) defining what the readers know; (3) defining what the readers need to know; and (4) defining what the readers will do with the information provided. Figure 3.1 depicts how these processes interact to determine the message to be communicated.

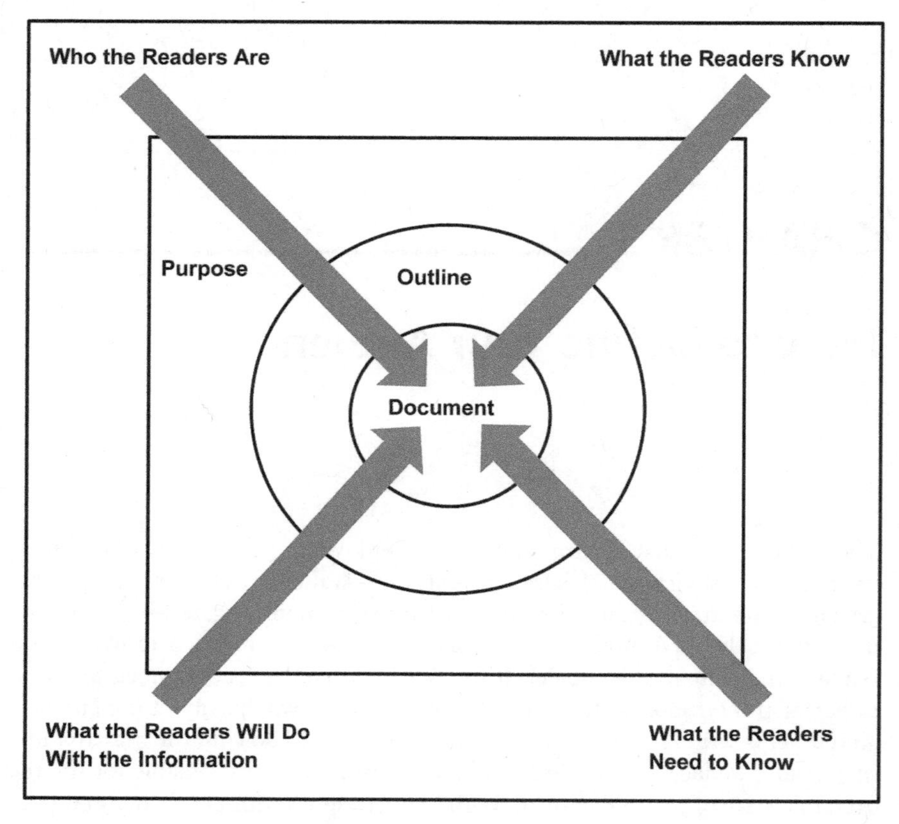

Figure 3.1. Audience Definition System.

Who Are the Readers?

If you are going to create and deliver a communication about a high-tech subject, the first thing you need to know is who your readers are. You need to know this information as specifically and in as much detail as possible. Because of the daily technological advances in high-tech industries, you are often writing in a vacuum without specific knowledge about your intended readers. Consider these examples: Are your readers managers who are responsible for the economic well-being of your organization, but who might not be fully versed on the technology your organization designs and develops? Are they engineers in firms that develop technological products? Financial planners in such firms? Students? Persons whom you've never met, but who regularly read the quasi-personal e-mail messages you send them daily as part of the professional responsibilities you both share? Or, is your audience made up of many of these groups?

For example, ZetaCorp, a hypothetical software company, has devoted its entire workforce to developing a new line of restaurant informatics systems. The project is to design a system that creates enormous amounts of information communicated electronically in real time across its customer base. As the project develops, however, its complexity is destroying the project schedule. Marketers who interface with the targeted customers discover that software engineers do not use the same vocabulary (jargon). The engineers find out that their reports are regularly sent to managers who have only a general idea of recent technological developments. The reports are abstracted for restaurant marketing and salespeople who are even less technologically informed. Everyone thinks someone else is screwing up, and the project falls farther and farther behind schedule. Sound familiar?

Readers come to a document with different backgrounds, different levels of knowledge about the subject, and different needs for the information that is provided. Management in firms might read articles on trade websites to learn to use their present systems more efficiently. Or they might find information that will suggest that they need to update or replace their systems. This information is especially important for small- or medium-sized organizations since a high-paced entrepreneurial attitude is required to keep companies profitable. As a result, management looks to the trade sites for trends and information that will affect their business for months to come. The writer's duty is to furnish this information in a form readers can understand and use . . . quickly.

Designers and software engineers must keep abreast of the rapid technological development in their fields, particularly since these developments signal competition. These readers make up a highly technical audience that needs highly technical information. Writers must meet the needs of these readers, also.

Financial managers in these firms do not have the same technical background as their engineers, but they need just as much information—with enough explanation so they can grasp the importance of new developments. They are concerned about competitors' developments because these developments might affect the company's future. Even rumor of development is important to these readers; they must keep ahead of trends to plan company marketing and development strategies.

Students might also come to a report or paper, especially in online resources that are available to the public. They might already know a considerable amount about the report's subject, or they might know virtually nothing. They come for learning and for information that will shape their careers. Articles in specialized business magazines furnish them with a great deal of this information.

And with the now almost ubiquitous worldwide reliance on e-mail and other forms of social media, we daily deal on an almost personal basis with colleagues we will never meet. We must plan for them, as well.

Of course, the worst situation is when projected audiences overlap. What can you do then? And how do you account for the different preferences readers have for how they comprehend and use information? It seems hopelessly complex (and social media has made it more so), but it is not. In such situations, your writing strategy depends on how much weight each component of your overall audience carries. Is it more important to focus on decision makers or end users and customers? On people who want step-by-step information or those readers seeking plenty of illustrations?

The following questions will help you identify prospective audiences. Your answers will provide you with information to begin making judgments about the type and quantity of information to include in reports or papers, as well as how to arrange that information. Later, this information will be your foundation for more in-depth considerations of readers' needs.

Who Is the Audience?

Try to answer this question in as much detail as you can. For in-house reports (including professional e-mail), even if these communications are transmitted to overseas subsidiaries, this step means identifying your readers by name and position. For papers and reports submitted for publication, it means finding out from the editors which type of reader is considered typical, or reading enough material in the publication to begin to develop a sense of that typical reader. Being able to do this sort of analysis helps bring writers out of the vacuum of trying to communicate to nameless, faceless readers. It is much easier to write if you have some definite ideas about for whom you are writing.

What Is the Educational Background of Your Audience?

With this question, you are trying to determine if the audience shares a common educational background with you. For example, how current is the reader's education? Is the reader's educational experience similar to yours? Ask yourself whether you could participate in a discussion of the topic with the reader on an equal level. These questions help refine your reader's identity.

What Is the Professional Background of Your Audience?

Here, you are trying to determine whether the audience shares similar experiences with you. This question is important because it enables you to determine what vocabulary to use, how much background is required, and which terms and concepts the audience needs defined.

These audience identification questions are important because they enable writers to get a picture of what the projected audience is like. Writing to an audience we can visualize is far easier than writing to an unidentified audience. Beyond that, identifying our audience forces us to start thinking about how our information should be organized.

What Does Your Audience Know about the Subject?

The first answer to this question is, it's hard to tell—without asking the readers personally, that is. The best way to discover what an audience knows is to do precisely that—ask them. But this requires very small, closed audiences—the sort that might exist in work groups who e-mail each other several times a day. In the past, the majority of communication environments precluded this possibility. Writers and readers were separated by space and time; they could not interface directly. E-mail and videoconferencing have changed this environment drastically, and in the 21st century, such audiences comprise the majority of the professional workforce in technological industries. What the writer needs in this situation is a method that is a little more systematic than guesswork.

Based on the answers to who the audience is, the following questions will help you determine what the audience knows. From the answers to these questions, you can begin to specify what you want the audience to know and what they need to know.

What Does the Audience Know about the Specific Topic of Your Report or Paper?

This differs from the main question in this section in that it focuses the possible answers. Even though an audience might have some general background knowledge of your topic, they might need considerable orientation to the specific topic you are reporting to them. For example, they might know a good deal about game-design architecture, but would need to be led clearly through a report or paper about the new type of virtual reality (VR) headset that your firm has developed. Be careful not to overassume the level of understanding of your audience. This assumption is the single easiest mistake to make in technical writing and the one that renders more information unusable than any other mistake. Even though your readers might share common educational and professional backgrounds with you, they have not stood by you while you developed the information for your communications. Nor can they read your mind. Don't make them try.

How Much Background Is Necessary?

This question must inevitably follow the first two. The answer to this question allows you to start thinking about the introduction of a report or the lead of a paper. For example, readers who share your educational background and professional experience will be bothered by long-winded introductions that tell them what they already know. Some research suggests that they will either skip over the irrelevant material or not read the communication at all, routing it to someone else. Certainly this reader reaction is not one of your goals as a writer. But the opposite of this situation is just as bothersome to another audience. Because as writers we tend to think our readers will automatically understand what we are trying to say, we are inclined to drop uninformed readers into the middle of specifics before we have told them the significance of the communication. This guarantees bewilderment.

These questions we have been considering are important because they begin to shape the information you have gathered about a topic of a report or paper. The process of audience analysis, incrementally focusing what you know about the people who will read your reports or papers, makes organizing information a simpler task.

What Does the Audience Need to Know?

This matter is subjective, for it depends on your purposes as a communicator. Is your purpose solely to provide information, as objectively as possible? Or do you intend to persuade readers to a particular point of view, perhaps even to buy a particular product? The answers to these two questions determine what you want your audience to know. In fact, the question is often best stated, what should your audience need to know?

A lot depends on your own professional situation. Are you working for a company that expects you to write an apparently objective communication, but also to display the benefits of a product they are developing? Such is often the case with papers published in various computer trade magazines; these documents have an obvious public relations or marketing bent to them. Or are you planning to give a talk at a professional convention or trade show? Here, the situation is even trickier. Your company will want you to reveal enough information so that the audience will understand the importance of late developments at your company, but you will not want to "give away the store."

The following set of questions will help you identify what your readers need to know.

What Information Does Your Audience Need from Your Report or Paper to Do Their Jobs Better?

The answer to this question forces you to start the process of examining all the information you have gathered about your topic so that you can delete what is not necessary. Too many writers feel compelled to jam all their material into a report or paper. After all, they worked hard to get that material; why throw it away? Editors in the trade press will make sure that irrelevant information does not stay in your papers, but you must ensure that it is deleted from your reports—even those that are distributed to industry websites, where the editorial function may not be more thorough than cut-and-paste.

Will the Information That Is Included Need a Technical Slant or a Managerial Slant, a Marketing or Public Relations Slant?

The answer to this question complements the answer to the questions that identify your audience. Even though you may explain your subject in ways that make it important to both management and technical readers, the communication should have a primary focus toward one group or the other. Depending on the purpose of the document, a marketing or public relations slant becomes either a primary or a secondary focus.

Sometimes writers are compelled by situations to give readers information they do not know they want or need. Accurate audience analysis enables you to provide this sort of guidance effectively when it is called for.

What Will Readers Do with the Information?

An important aspect of deciding what information you will include in communications is determining what readers will do with it. Knowing the answer to this question enables you to avoid including information no one needs and cluttering the report needlessly. The following questions will help you identify what readers are likely to do with the information you provide them.

Will Readers Use the Information in Your Communications to Perform Professional Tasks?

The answer to this question helps you to decide if your report or paper should have a procedural organization. Will readers need clearly identified steps to follow, as in a manual?

Will Readers Use the Information in Your Communication to Increase Their Knowledge of the Field?

Everything from formal analysis reports to e-mail messages can fall into this category. The answer to this question helps you to decide if the audience needs to see how the information in your report or paper fits into the literature and practices of their field.

Only by visualizing what readers will do with the information you plan to give them can you be sure that it is the right information, accurately communicated.

What Personality Types Will Be Reading Your Report or Paper?

Keep in mind that different personality preferences affect audience needs for information, as well as their reading strategies. If you are writing for one or two people whose preferences you know, then you can gear your presentation to their type. This is particularly important for e-mail messages; think of them as typed conversations. In larger communications environments, you will have to try to include information for the main categories of types in your audience. For example, intuitives generally need just enough information to get started. Overviews are important for them to get the sense of what they will be reading, as are information and document design techniques that enable them to skim a document. Sensing types will want more detailed information, meeting their needs for step-by-step analyses or procedures. Extraverts will often use your information verbally, usually in meetings. Introverts may file it for use in other documents. Whatever the situation, you need to think about these differences and the roles they play in your writing.

CONCLUSION

This chapter has presented a system for taking general information or gut feelings about the reader and turning them into specific details that you can use to help organize information for a report or a paper. In a sense, the method is subversive; it treats the matter of audience definition from the stance of giving the audience what it wants. This is a variation of the philosophy that the customer is always right. But all people who have ever had a college course in which they have attempted to write what the instructor wanted probably figured out the importance of this philosophy somewhere during the course. We neglect to realize, however, that the same situation exists in professional communication. We might not always tell readers what they want to hear, but we should always give them what they need—and should want—to know.

The following decision-logic table summarizes the processes presented in this chapter (Figure 3.2). It does precisely what its name suggests: provides information with which decisions can be made logically. To use this one as a checklist for defining audiences, match the conditions that you believe are present for the document you intend to write (*Y*=Yes, *N*=No). In this table, six combinations of conditions are possible. Depending on how the conditions you have identified for your report match up with those listed in the table, follow the appropriate actions marked with Xs.

For example, if you decide that you are addressing a management audience only and that your purpose is only to inform (conditions common to many short, printed reports and e-mails), you would focus the document on matters that affected budget and personnel matters, schedules, or decisive actions that need to be taken, including the specific facts and supporting data as necessary. Depending on the needs of the individuals you are addressing, however, you might decide that some design information is necessary (e.g., addressing a design project manager). That is okay; trust your audience analysis. Your primary focus for the information will probably still be on matters that show how the design information fits into the overall project plan and schedule.

		1	2	3	4	5	6	
Conditions	Management Audience	Y	Y	Y	Y	N	N	Else
	Technical Audience	Y	Y	N	N	Y	Y	
	Informative Purpose	Y	N	Y	N	Y	N	
	Persuasive Purpose	N	Y	N	Y	N	Y	
Actions	Include Budgetary & Personnel Information	X	X	X	--	--	--	--
	Include Specific Design Information	X	X	--	--	X	X	--
	Include "Facts & Figures"	X	X	X	X	X	X	--
	Include Common Ground Orientation Between Reader and Writer	--	X	--	X	--	X	--
	Rethink Report	--	--	--	--	--	--	X

Figure 3.2. Decision Logic Table for Audience Definition.

And remember, different people read and interpret information in different ways. Provide several paths of access to important information.

SUGGESTED READINGS

https://owl.english.purdue.edu/owl/resource/629/1/, May 25, 2017.
http://www.wikihow.com/Conduct-Audience-Analysis, May 25, 2017.

PART II

Getting Started

CHAPTER 4

How to Get Organized

Every time I teach a communications seminar, I am inevitably told the same thing by course participants: "The most difficult thing for me to do is to get started." This statement is true for almost everyone. I don't think I have met a single person, including people who make their living by writing, who have said that they can sit down anytime, anywhere, and start writing. The source of this problem is getting organized, and if it is your problem, you are not unique. Everyone experiences the problem of getting organized.

For some of us, it means an almost debilitating situation in which we will put off starting a document, using any excuse that is handy. "I don't have time right now because I am busy," is the most popular reason for not writing. But even, "I can't start today because I am having a root canal," will suffice. For other people, getting started is only a hurdle that must be crossed before successful writing can take place. These writers, many of them professionals, have solved the problem by developing personal systems for getting started. Although the systems differ from writer to writer, they have one thing in common: they force writers to work by providing them with successful ways of getting started.

In this chapter, we will look at one method for beginning a paper or report. It has worked for a lot of people, even for those whose personal work-style preferences might not match well with this system. And although I can't guarantee it, it will probably work for you—if you follow it rigorously.

Figure 4.1 introduces a concept I call Visual Organization Systems (VOS). Taken from flow-charting techniques, these visuals will be used throughout the book to summarize procedures discussed in various chapters. The VOS depicted below outlines one approach to writing; each subsequent VOS, which

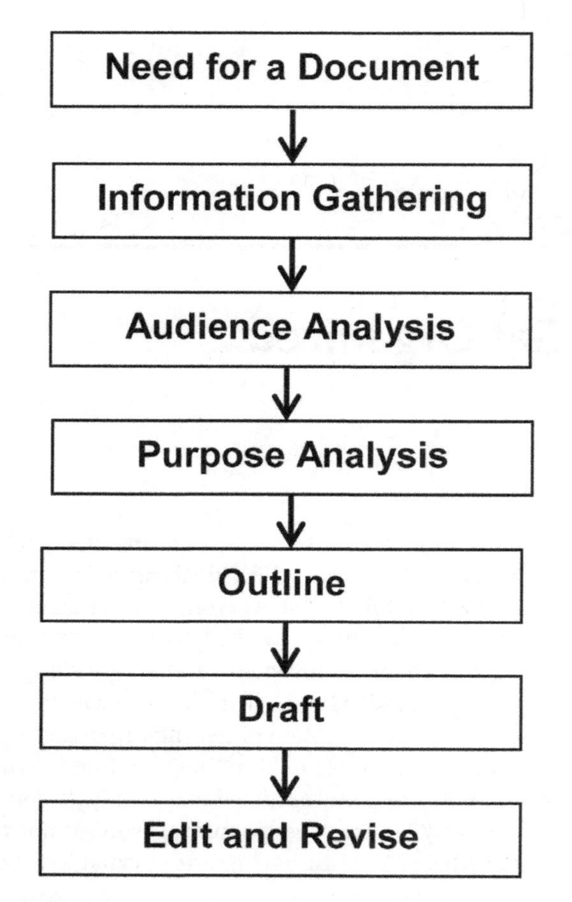

Figure 4.1. VOS for Writing.

deals with a particular type of document discussed in this book, would be plugged into the outline box in Figure 4.1.

AUDIENCE ANALYSIS

The first matter over which you have some control is audience analysis. If you are lucky, this analysis has been started by whoever assigned you the communication task in the first place. For example, you may have been told by a superior to write a report about a particular project and to address it to a particular person or group. If so, continue the audience analysis along the lines discussed in Chapter 3. If you have initiated the communication—or worse, if your superior has given you no direction as to who the audience is—then audience analysis is even more necessary, and it should begin with the first step described in the previous chapter. Before you can identify what the audience knows, you have to know who the audience is.

PURPOSE ANALYSIS

The second matter to consider is the purpose of the project. In fact, you can think about it while you are doing audience analysis. Doing so is probably unavoidable. But since most of us cannot do two things at once, it is helpful to analyze your purpose immediately after you have analyzed your audience.

Realize that every document has a purpose. A helpful way to make yourself think about this is to write out the following sentence before you write anything in the document itself:

> *The purpose of this document is to* ___________.

Then fill in the blank. For example, it might look like this:

> *The purpose of this document is to provide technical updates to the software design group for Project XYZ.*

This simple exercise should lead you to identify three important points about your planned document:

- The problem the document addresses
- The technical issues or major points to be made
- The rhetorical issues or what the document will do for readers

If you cannot identify these things, don't start writing. You are not ready. You need to gather more information about your subject, the situation, or the audience.

After you have analyzed your purpose, the best way to use the results is to design the document's outline around them. If you do, outlining becomes a simple and painless task. Also, the purpose analysis can be used as an introduction to most reports. In fact, it is the best type of introduction for most reports because it orients readers to the central issues in a document and tells them specifically how the document will meet their needs.

The following example is an introduction from an analysis report. What it shows is the benefit of providing even highly technical audiences with the types of information listed above. Although it is not necessary to divide an introduction into three paragraphs (each for a different aspect of purpose analysis), notice that the following introduction does in fact treat each of those aspects:

> Any design process begins with a clear definition of the task. In the case of a microprocessor, the task involves either controlling something or manipulating data for the means of achieving an end. The microprocessor design detailed in this report is of the second type—it is to search through 512 locations in a video game's state memory and generate the data and control signals to display the video game in its present state on an x-y vector display.

The state memory is arranged such that the data blocks for each body to be drawn are in permanently assigned locations in the lower half of the memory. The second half is set aside as a character list: sequential data blocks identify selections from a list of 38 possible characters to be displayed on the screen where specified by the coordinates in the data block.

This video display unit requires two clock signals from the video game state control or "update" unit. In addition to the required access to the address lines and the data lines of the state memory, the update unit must supply: (1) a clock that signals which processor is to control the state memory and (2) a common 1.33 MHZ system clock for easier processor synchronization. The state memory control clock (V clock) must change state every 214 system clocks, giving each unit 12.3 ms to perform its task. This restriction is necessary, because allowing more time results in game displays happening slower than 40 times per second, the rate at which screen flicker becomes noticeable to the eye.

To provide a more precise task description, I will describe the video game and how it is to appear, followed by a map of state memory and the information contained therein. Following this, I will develop the system architecture designed to perform the appointed task. The result of the architecture design is the instruction set with which the program can be compiled and the specifications with which the hardware can be built. I will discuss the microprogram as a way of summarizing the architecture; then I will conclude with an outlined description of the realization and debug of the prototype.

INFORMATION GATHERING

Gathering information is the lifeblood of writing documents. Information is what readers want; delivering it to them is the writer's primary purpose. The introduction above points out how much can be done for readers when writers have sufficient information about audience, purpose, and subject.

Once readers understand the points from an introduction, they are appropriately oriented to the details that follow.

Information may be gathered by a number of methods, including observation, experimentation, literature, and online searches. Another method for gathering information, which is not considered as often as it should be, is through fact-finding interviews. Interviewing is particularly appropriate for people writing about high-tech because developments occur at such a rapid pace that the literature rarely keeps abreast of them. The best way to find information, then, is to ask for it. Chapter 5 will explore how to conduct those interviews.

OUTLINING

Outlining is something everyone should do before starting a document. Now, even though this sounds like typical academic advice, it isn't. Outlines have a purpose, beyond busy work for active junior high school students. They provide writers with a framework for the document. Doing so, they also provide the first indication of whether the intended purpose is being met for the intended audience. If the purpose is not being met, it takes far less time to rewrite an outline than it does to rewrite an entire document. For this reason alone, using outlines makes sense.

The type of outline you use is not important. This statement might be heresy to some English teachers, but it is true in the real world. An outline is only a device to help writers get organized, nothing more. Any formality, or lack of it, in an outline is strictly the business of writers; whatever they are comfortable using and whatever succeeds for them is right.

For example, outlines may be formal sentence outlines in which the topic sentences of each paragraph of every section of a report or paper are written out. This type of outline makes writing the rough draft of a document fairly easy. But formal sentence outlines have disadvantages as well. Their rigidity or formality makes it less likely that editing will be done as rigorously as it should be. Writers who write a draft from a formal sentence outline are reluctant to change things in the editing phase—even when things should be changed. The rough draft looks better than it is.

On the other hand, writers may opt for informal phrase or note outlines. The purpose of these outlines is only to arrange ideas throughout the report or paper. No attempt is made to actually write the ideas out during the outline phase; this step is saved for the rough draft. The advantage to this outlining method is that the rough draft will be rough. And that's as it should be. A rough draft promotes effective editing and improves the final document. It even decreases the writer's overall work (although the opposite may seem to be the case) if we factor in the time spent rewriting a document after it has been sent out and returned because it did not work.

In addition to these common approaches to outlines, some writers (myself included) may even prefer mind maps (visual or picture outlines). Figure 4.2 is an example that I used as the basis of a 90-minute presentation I made as a consultant to a well-known computer company.

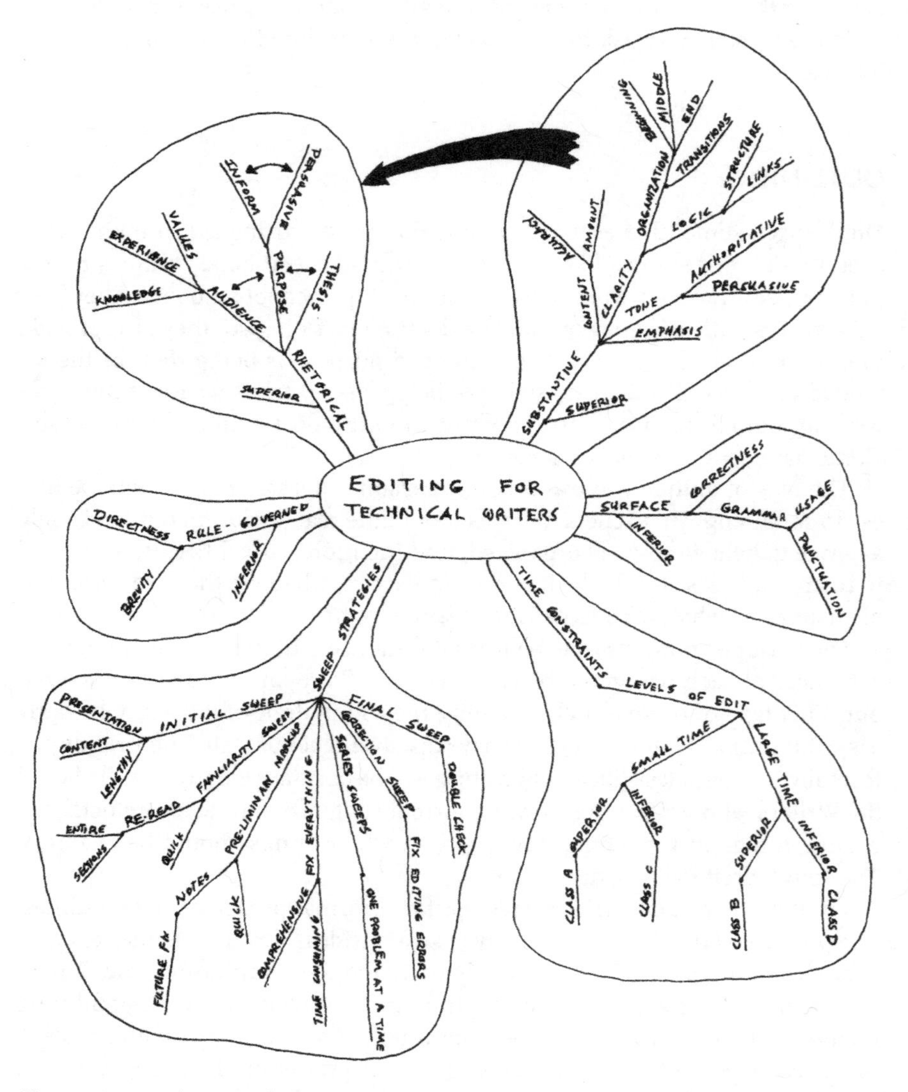

Figure 4.2. Mind Map Outline.

As strange as an outline such as this one may appear, it works particularly well for people who prefer to keep writing (or speaking) projects open-ended, people who fall into the intuitive-perceptive categories mentioned in Chapter 1. Sensing-judging types tend to prefer outlines that are more formal and structured. Just remember: there is no one correct way to outline. Use what works for you.

To help you determine what method works for you, notice that outlines appear to be inherent in some subjects. For example, if I were to outline a physical description of the MacBook Pro laptop I am currently using, I immediately notice that I am limited by the subject. The machine, as I have configured it, has four physical parts: the laptop itself, a 20-inch display screen wired to one of the laptop ports, a Bluetooth keyboard, and a Bluetooth mouse. The only thing I have to do is to decide in what order to describe them. The order I listed in the previous sentence is a natural one—the order in which most people would notice the parts of my computer if they were sitting in front of it. The laptop is prominently positioned on my desk. Next, a person would almost certainly notice the large display beside it. After that, the eyes logically move down to the keyboard in front of the display. And finally, off to the side of the keyboard, is the mouse. So, at this stage, the outline might look like the following one:

OUTLINE: MacBook Pro Laptop—PHYSICAL DESCRIPTION

1. Laptop
2. 20-inch display
3. Bluetooth keyboard
4. Bluetooth mouse

The next thing I might do to make the outline more useful is to organize my description of each individual part. This step would give me a two-level outline and a good deal more information about my subject. The laptop is a gray plastic case, with a flip top, which when open reveals the display, the on/off switch, a keyboard, and a finger pad. The keyboard has the standard arrangement of typewriter keys, a control keypad, an arrow keypad, and a row of function keys. The display unit is a 20-inch monitor mounted on a silver plastic pedestal. The Bluetooth keyboard is a standard keyboard with arrow keys, a row of function keys, and a number-key section. The Bluetooth mouse has a prominent trackball on its top. Notice in these lists, however, that I have again followed a first-noticed organization plan. Now my outline might look like this:

OUTLINE: MacBook Pro Laptop—PHYSICAL DESCRIPTION

1. Laptop
    1.1 Gray plastic case
    1.2 Built-in 13-inch display

    1.3 On/off switch
    1.4 Built-in keyboard
2. Display Unit
    2.1 20-inch color screen
    2.2 Silver plastic pedestal
3. Bluetooth Keyboard
    3.1 Standard typewriter QWERTY keys
    3.2 Row of function keys
    3.3 Arrow keys
    3.4 Number keys
4. Bluetooth Mouse
    4.1 Beige plastic case
    4.2 Trackball

If I were to go on, I would notice that several of these second-level topics could be further subdivided into third-level topics, and so on.

By now, however, you should have noticed several important things about outlining. First, the order of the outline depends upon what you decide is important about the subject, and this decision depends upon your analysis of the audience's needs. For example, if I had chosen to do a functional description rather than a physical description, my outline would have been necessarily different. Second, if you are familiar with the MacBook Pro laptop, you know that I was outlining my description from the point of view of standing in front of each component. As a writer, you have to make some decisions about point of view, and that depends on your purpose. Remember: The material you include in a document and the way you arrange it depend ultimately on your purpose in meeting the reader's needs.

ROUGH DRAFT

The first draft of a document should be quick and dirty—in a word, rough. The goal is to get ideas into writing as quickly as possible, leaving editing for another time.

Remember my comment about people not being able to do two things well at the same time. If you try to write and edit at the same time, you will do neither well. But avoiding editing while writing takes tremendous willpower. If you use a computer for your drafts, that delete key is a temptation. Try to avoid it; it slows your thought processes down and forces them to be two-directional—forward and backward. Also, the result of this type of writing is not as good as it should be. You can print the rough draft on a laser printer, and it will come out looking great. You'll be tempted not to change a thing, and then the rough draft problems, which will naturally be there, are passed on to your readers.

A successful rough draft is messy, and the writer should be the only person capable of fully understanding it. After the rough draft is complete, leave it alone for 24–48 hours. Research has shown that we need this much time to forget what we wrote. After that time, we can come back to a draft and edit it from the reader's perspective, correcting problems in logic and organization. If, on the other hand, we write and edit at the same time, we will still have all the ideas floating around in our heads; we will subconsciously make connections in logic that are not actually there; and we will be too impressed by our own brilliance. In short, we will not change what we should, and the reader will suffer for it.

One final problem often encountered during the rough draft is "writing roadblocks." These are dealt with specifically in a later chapter. But more than likely, if you find yourself blocked during the rough draft, you are probably not writing a rough draft. Rather, you are trying to write and edit at the same time. Don't.

EDITING

Editing, which requires locating the weaknesses in a rough draft, and the subsequent redrafting of a document, makes up the final stage of this approach to writing. Methods for doing this are fully discussed later, in Chapter 25, but the process should be briefly mentioned here. Editing should be an organized activity. First, edit for problems of logic, organization, and clarity. Can you (or someone else) follow the development of ideas through your document? Next, edit for style. Are the sentences varied in length and structure? Are they interesting? Finally, edit for grammar and spelling. Too many poorly written documents are the result of editing done in the wrong order. Often writers get overly concerned about the nit-picking details of grammar and punctuation in a document before they have got the thing written clearly. Once it is written clearly, then pay attention to these details. And don't trust grammar or spell checkers either. Spell checkers have improved greatly over the past 30 years, but there are some things that they are not even close to being able to do. For example, if you write "The book was read," but you meant to write "The book was red," there is not a spell checker on the planet that can identify which statement is correct. Grammar checkers are useful only for entertainment value. More often than you might guess, they are wrong; and worse, their errors limit the style and creativity that writers can bring to bear, even in technical documents.

In the end, good editing enables writers to discover areas where the purpose is not being met, areas where the development of ideas is not lucid, and at last, areas where grammar, punctuation, and spelling are faulty. Don't overlook editing.

CONCLUSION

In this chapter, we have examined one process of writing from start to finish as a way of getting organized. It is not the only way, but it is a successful way. Following these steps for every document you write will come as close as anything you can do to ensuring that your documents succeed.

SUGGESTED READINGS

Strunk, William, and E. B. White. *Elements of Style*. Value Classic Reprints, 2016.
Zinsser, William. *On Writing Well*. New York: Harper Perennial, 2016.

CHAPTER 5

How to Get Information with Interviews

Information for documents can be obtained in many ways. The most common are observation, experimentation, and reading. Observation is a valuable tool for gathering information because it is firsthand experience. The writer is present at the event that is being reported. The disadvantage with observation, however, is that it is totally dependent upon the viewpoint of the observer. Such viewpoints are rarely accurate, they are always subjective, and they vary widely from observer to observer. A perfect example of this phenomenon is how witnesses to accidents consistently give different accounts of what happened. Or, consider this: Years ago, I had the privilege of taking a graduate course with a famous linguist. One day, she walked into the amphitheater where approximately 50 graduate students awaited her brilliance. Saying nothing, she went to the center table, which was visible to all, took two rubber balls—one large, one small—from her pocket, set them on the table, rolled them toward each other, where they hit, resulting in both falling off the table to the floor. Then she said, "What did you see?" The range of observations was astonishing, varying from relatively benign descriptions of the events to imputing sinister motives on the part of the large, aggressive ball. Observation is a wonderful tool, if we recognize and account for its challenges.

Experimentation is also firsthand experience, but it is more accurate because its procedures are rigorously controlled. Information generated by experimentation is usually considered to be the most valid type. Design processes, and the information they generate as a by-product, are a combination of experimentation and observation. Common to the computer industry and other engineering fields, these design processes normally consist of creating prototype products and then testing them to determine how well they perform. Consequently, this type of information is considered reliable, too.

Finally, reading can be used to generate information. Reading is second-hand experience, though, and its accuracy is highly dependent on the accuracy of the original writer. In the 21st century, when Wikipedia is a major resource and fake news is widespread, be certain that what you read is valid, well supported, and reliable.

But another way also exists to obtain useful information—interviews. When we stop to think about it, we use this technique more often than we realize. A telephone conversation that seeks information is a type of interview. So is a face-to-face conversation. Much of the give-and-take that we participate in via the Internet can also be thought of as a type of fact-finding interview process. The problem is that we rarely think about the interview aspect when we talk with someone else on the phone or face-to-face, or when we jot down quickly sent e-mail requests, and we almost never plan our conversations ahead of time.

Well-planned interviews, however, can be a valuable source of information for the person who has to write about high-tech subjects. In this chapter, we will look at one way of designing and using them.

DESIGNING AND PLANNING INTERVIEWS

Designing interviews requires us to plan them and to anticipate the reactions and answers of our interviewees. The central point to keep in mind is that we are searching for important information, and everything must be structured. As we have already seen, however, people differ in how they communicate and in how they respond to communication. As a result, audience analysis is important in designing an interview. You need to know your interviewees (or at least something about them). How receptive are they to being interviewed? Are they outwardly open or not? These issues, and others, determine how successful your interviews will be.

Remember that in-person interviews are extraverted experiences. Some people like them and others do not. You will want to take this fact into account when planning for and designing an information-gathering interview. For example, conducting an interview with a person who is forthcoming is easy. You basically plan a conversation to elicit information you need. If either or both the interviewer and the interviewee are introverted types, however, the process could be very difficult. But with careful planning, the interview could still be completed successfully. In these cases, you will want to prepare open-ended questions to draw out needed information.

E-mail questionnaires, on the other hand, are introverted experiences for both the questioner and the respondent. Nonetheless, they still need the forethought and planning we associate with any type of interview. And regularly, this planning does not happen when people use the Internet to obtain information. Avoid slapdash e-mail queries. They confuse the respondent and almost never provide the information we sought—at least not on the first try.

Preparing for an interview goes beyond scheduling a time that is convenient for you and the interviewees. The most important thing you can do is spend some time learning about the subject of the interview. If this sounds like a variant of audience analysis, it is. Nothing will doom an interview more quickly than if the interviewees think you are wasting their time. And the way not to waste interviewees' time is to show that you have done your homework. People are much more likely to share information with you if they are convinced that you have taken the time to learn something about their specialty. Reading is the best way to brief yourself about the interview. It is even more effective if you can read something the interviewee has written; this type of flattery never hurts.

Opening Remarks

Whether you are conducting an interview over the telephone or face-to-face, the first thing you have to do is convince the interviewee that your need for information is important and worth the time it will take to meet it. Introducing yourself and the company or division or group you represent are the bare minimums. After you have introduced yourself, it is vital to get to the point of the interview immediately by stating what the problem is—why you need information. This explanation gives your interviewee a rationale for being helpful.

Problem Background

Explain enough background to the problem so that it orients the interviewee to your mission. This step requires you to explore two areas: how the problem can be defined and how it was discovered. In other words, give the interviewee a clear understanding of the problem and its ramifications before asking for specific information. But remember: Keep this aspect of the interview short; it's still the introduction.

Interviewee Incentives

If additional convincing is necessary at this point, it is a good idea to tell the interviewee why participating in this process is beneficial. If it is not beneficial to the interviewee, then you should rethink the situation. It ought to be. Rethinking usually entails only redefining the problem or re-evaluating your mission. At any rate, you must anticipate all this while planning the interview. Don't allow it to become evident during the interview.

Direct Request for Help

After the introductory remarks, ask the interviewee directly for assistance. Give the person an idea of how much time the interview will require. If you have planned well, "yes" is the only possible answer—you know your subject and situation that well. Remember, however, that all this conversation is only preliminary to the questions that will get the information you need. Don't forget to keep all introductory parts of the interview as short as possible—a couple of minutes at most.

Information-Gathering Questions

To obtain the information you need, go into an interview with a list of prepared questions. The list will probably contain a combination of open-ended questions and specific questions. The open-ended questions are designed to get interviewees to elaborate on a subject, to share their range of experiences with you. The specific questions are aimed at gathering precise information.

But even with open-ended questions, make sure you have designed them to get results. "Open-ended" is not a synonym for "unplanned." Don't say: "Can you tell me something about the new operating system?" A planned open-ended question would be more like, "I understand that the new operating system is designed to do x-y-z; could you elaborate on that?" If, on the other hand, you are looking for specific information, your questions might be more like, "How will the new operating system improve information flow across our organization?" or "How does the new operating system compare with other operating systems in its ability to manage data quickly and efficiently?" In other words, whether your questions are open-ended or specific, lead the interviewee toward information that you can use. Prepare enough questions to cover the subject, and plan for follow-up questions, too.

Three Structural Alternatives

Interview experts describe three alternatives to structuring the body of interviews: loosely scheduled, moderately scheduled, and highly scheduled. Loosely scheduled interviews occur when the interviewer organizes the interview with a list of topics to be covered and then moves about somewhat randomly within that list. Such an interview is organic; it develops its own structure as it is occurring. The advantage to such a structure is that it is dynamic, and as a result, it can lead to new, unthought-of, and potentially important areas of information. The disadvantage is that it can appear completely disorganized and a waste of time to a busy interviewee.

Moderately scheduled interviews are those that I based much of this chapter on. The interviewer prepares a list of questions and the order in which those questions will be asked, along with some potential follow-up questions, if necessary. This is the most common type of fact-finding interview, and it consistently yields much information in a minimum amount of time.

Highly scheduled interviews are most often found in marketing analysis or similar information-gathering endeavors. The interviewer has a complete list of questions, possible answers, and follow-up questions based on those answers. This type of interview is good for what it is designed for—usually the revelation of attitudes and opinions about products or potential products, but it is a bit cumbersome in other applications. If you have ever said "yes" to a person in a shopping mall who asks you if you have a few minutes to answer some questions about XYZ product or service, odds are that you have participated in a highly scheduled interview.

Closing Remarks

At the close of an interview, let the interviewee know how important the information is (again!). Don't forget to express your appreciation for their time and cooperation. You might even add that you will inform the interviewee about the results of your communication project. And you might ask whether a follow-up interview could be done if necessary.

Focus Groups

An additional useful interpersonal method of gathering information is focus groups. Think of these as group interviews. Sometimes called facilitated group discussions, focus groups can be an excellent method for gathering information, as well as to determine specific methods to deal with issues, come up with group decisions, and so on. The challenge to focus groups is that, unless they are highly developed and skillfully run, participants can view them as a huge waste of time and the results will be nearly worthless. Unfortunately, too many businesses these days run such types of focus groups, and as a result, people who are forced by their managers to participate see them as unproductive intrusions on their work.

But skillfully conducted, they can be very useful. Let's examine how to do this.

A focus group is usually an interaction among 3–10 participants who are guided by a facilitator who manages the interaction and guides the discussion toward predetermined objectives (what you want to accomplish in the discussion). The facilitator's responsibility is to ask probing questions and provide a nonthreatening environment for the exchange of ideas and opinions. That last part is very important. Focus groups succeed only when participants feel they will not be judged or harmed for the information they share with the group.

There are several advantages to focus groups. They involve more persons (and consequently can tap into more than one viewpoint at a time) than interviews do. As a result, focus groups permit people to learn from others' experiences, attitudes, knowledge, beliefs, and values. Finally, done right, focus groups are safe for the participants.

But there are challenges, too. Facilitators must be able to keep the discussion focused on predetermined objectives. Otherwise, focus groups can degenerate into pointless gabfests. Facilitators also must be able to draw out participation from everyone in the group, shy people and outgoing people alike. And, depending upon the topic of discussion, facilitators must be adept at managing emotions that might arise in focus groups. This last challenge might make focus groups especially tricky to use for inexperienced persons.

However, if you decide to use focus groups as a primary research tool, there are techniques to keep in mind and use.

You want to expose participants to some sort of stimulus to get the conversation going. This might be an open-ended, or "trigger question," that you display using PowerPoint, a chart, a board, etc. Then you might ask them to think about that question for a period of time (no more than a couple of minutes) or you might want them to jot down their ideas during that time, as well.

Next, you want to set the boundaries to the discussion. These might include not interrupting others, remaining respectful, describing rather than evaluating, self-disclosing how other opinions make one feel, asking questions when disagreeing rather than verbally challenging another person's opinion, pointing out that one can understand another opinion without agreeing with it, etc.

Then, based on the rules, you begin the discussion. Ask additional questions. Encourage all to participate. Provide summaries of what transpires. Redirect the conversation if you sense it is going off-topic. Monitor nonverbal behavior to make sure everyone is engaged, unthreatened, etc. Draw out minority opinions if you sense they exist. Tactfully, shut down persons who monopolize the conversation.

Done right, your role as facilitator of focus groups can amaze you with the quantity and quality of the information you discover.

CONCLUSION

In this chapter we have looked at an alternative method for gathering information for reports and papers about high technology. Interviewing is a valuable tool because many high-tech industries are extremely compartmentalized. In large companies, hardware people may not know what software people are doing; programmers may not know what technicians are doing; and no one in the technical areas knows what the marketing and salespeople are doing. In addition, information is produced as fast as new products are produced, and new products are produced daily. As a result, reading gives you hopelessly outdated information, although web pages, depending on how often they are updated, have ameliorated this problem somewhat. Observation requires you to be all places at once. Experimentation requires you to know something about everything. That leaves only interviewing as a consistently successful method of getting useful information. Incorporate it into your bag of writing tools.

SUGGESTED READINGS

Brinkmann, Svend. *Interviews: Learning the Craft of Qualitative Research Interviewing*, 3rd ed. Los Angeles: Sage, 2014.

CHAPTER 6

How to Explain Your Subject

Explaining oneself is the craft of clear communication. It can all be reduced to this common denominator: rehtoric. Rhetoric, in the apolitical sense, is the body of techniques by which we explain our knowledge of a subject to an audience. Nothing about it is new; most of the ideas in this chapter have a 2,400-year successful history. Nor is any of it mysterious or particularly complicated. In fact, most of it is common sense. Moreover, not all aspects of rhetoric are applicable to technical communication, but many are. Among the most useful are the following techniques: definition, classification/partition, comparison/contrast, cause and effect, and deduction and induction.

DEFINITION

Six types of definition techniques occur in documentation: formal/informal, operational, metaphorical, contextual, stipulatory, and divisional.

Formal/Informal Definitions

Formal definitions are the most common type of definitions found in documentation. In fact, they are common to all types of communication in which the purpose is to explain something new to audiences. Correctly written, formal definitions adhere to a rigid format:

> *Term to be defined = class to which it belongs + differentia (information that distinguishes the term from all other members of its class)*

Example: Artificial intelligence is a complex aspect of computer programming that strives to model human intelligence.

When designing a formal definition, remember that the = almost always is some form of the verb "to be," and that the differentia is the information following the word "that."

Formal definitions are most often found as the topic sentences of paragraphs, or as the lead to a new segment of information within longer sections of reports. They establish the subject and help to expand or explain it. When writers choose to expand formal definitions into product descriptions, they commonly use examples, illustrations, and further definitions. If the topic being described is complex enough, or if the treatment of it is involved enough, formal definitions may be used to begin sections, papers, reports, or even books. For an example of how this works, look again at the first paragraph of the Preface.

Informal definitions are actually formal definitions that have been altered for the sake of stylistic variety. Often, if you are describing a complex device, piece of equipment, or product, you would not want to begin the description of each component with the same communication technique—a formal definition. By using informal definitions in some of these instances, you can vary your sentence structure, and by doing that, you can keep readers awake. The following example shows how short, formal definitions can be combined effectively into a longer, informal definition:

1. *An ohmmeter is an indicating instrument that directly measures resistance in an electric circuit.*
2. *An ohmmeter was used to test the circuits on the IC board.*
3. *An ohmmeter, an indicating instrument that directly measures resistance in an electric circuit, was used to test the circuits on the IC board.*

Notice that the third sentence embeds the formal definition of sentence one into the procedural statement of sentence two. The result is a more informative sentence, as well as one that is more stylistically interesting. Informal definitions are one way to avoid an overly simple writing style while still maintaining logical clarity.

Operational Definitions

Operational definitions are most useful in describing processes or procedures. They do so by explaining how the process changes over time or how it works, or by describing a way to measure it. For example:

Death has occurred when an electroencephalograph reads a flat wave pattern for a predetermined period of time.

Notice that the format for an operational definition is not as rigid as the format for a formal definition. Rather than using a "to be" verb that establishes the state of something (as in a formal definition), operational definitions use action verbs. If the writer's purpose had not called for an operational definition, a formal definition would have been just as easy to use. But notice how it changes the sense of the information:

Death is a physical state in which . . .

Using formal definitions in procedural writing, though allowable, almost always results in wordier, weightier writing.

Metaphorical Definitions

Many people who lack imagination or the abilities of close observation might insist that metaphors have no place in technical communication, that they should be consigned for all time to the nether regions of liberal arts or creative writing. There are a few problems with such an attitude. First, all writing is creative in that a human mind is selecting words to communicate an idea to another human mind. Second, the belief that metaphors have no place in technical communication is blatantly ignorant of reality, particularly insofar as communication within the high-tech industries is concerned. Metaphors are everywhere. True, many of them are so overused that they have become trite, worn out, and unnoticeable, but they are there; high-tech industries have an especially rich heritage of dreaming up metaphors and foisting them upon everyday communication. For example, who would think of "bug" in an entomological sense? If someone said, "I have to debug this program," would anyone rush to call an exterminator? Of course not.

In fact, the history of this particular metaphor in the computer industry is interesting. Early in my career, when I was on the faculty at the Massachusetts Institute of Technology (MIT), I served on a committee that also happened to have a distinguished member of the computer engineering faculty, who had begun his academic career at the close of World War II. We happened to arrive at a meeting a bit early, so we started chatting. He said, "Charles, have you ever heard the background to the term 'a bug in the system'?" "No" was my response. And then, he told me the following story:

Back in the early days of computer research, when he was a new faculty member at MIT, a team was constructing one of the first prototype computers;

it was the size of a room, filled with vacuum tubes, wires, and other heat-producing gadgets that failed pretty regularly. All this to process simple data, such as adding, subtracting, etc. As usual, it crashed one afternoon, and the team sent a lowly grad student into the computer through a small access door to find the source of the problem. A few minutes later, he emerged with a fried roach held in a pair of tweezers. There truly was a bug in the system.

Now, this story may be apocryphal, but a few years later, I was at a professional conference in which Rear Admiral Grace Hopper, the ranking woman in the U.S. Navy, was giving a speech, and she told the exact same story, except that she placed it at the University of Pennsylvania, another research institute at which the first prototype computers were being built. I suspect Cal Tech (California Institute of Technology) has a similar myth.

But the point to all this is that metaphors are richly woven into technological language. Consider today: We no longer talk about bugs, but every one of us has said something along the lines of "I think my laptop has a virus." I have even seen products described as having the capability to inoculate one's computer against viruses, thus extending the metaphor.

Despite their rich history, metaphorical definitions are not methods that writers should be avidly hunting, however. To do so would be a waste of the writer's time and not very successful in the long run. Good metaphors tend to evolve more often than they are created. And they start in everyday, on-the-job conversation. From there, they become the jargon of a limited group. If they prove to have wider applicability, their use expands to larger and larger groups until the general public uses them and they have lost most, if not all, of their original usefulness. But such is the dynamic nature of the English language; it's what makes communicating fun and challenging. Metaphors (and their understood definitions) create much of what we identify as industry jargon.

Contextual Definitions

"Contextual definitions" are basically truncated formal definitions. The important thing is that the definition depends on the context in which a term is being used. For example, take the word "bond." In the context of psychology, its meaning has to do with relationships. In mechanical engineering, it is a fixative, a glue perhaps. In finance, it is a type of investment. In chemistry, it has another meaning, in physics another, and so on.

The use of contextual definitions depends on the sophistication of your audience. Writers must always be sure that readers are aware of the proper meaning of terms within the communication's context. That awareness often determines the extent of definition needed in a document, from a glossary to simple in-text definitions.

Stipulative Definitions

"Stipulative definitions" are a narrower type of contextual definition, used within the context of single documents or document sets. For example:

> *In this report, the term "light sensitive liquid resin" will always refer to the medium used in stereolithography for 3D printing.*

The advantage of using a stipulative definition is that the writer can preset the context and the meanings of terms when they are first encountered in a document. Used this way, an alphabetized list of stipulative definitions is a glossary. If the glossary is placed at the beginning of a report, the audience will have been informed of the specific context in which to read a document. As a result, stipulative definitions are most useful when writing to audiences that do not share the writer's knowledge.

Divisional Definitions

"Divisional definitions" are not really definitions at all; they are organizational techniques that can be used to expand other types of definitions, most often formal definitions. Usually, divisional definitions set up extensions of a subject, the kind often found in technical descriptions. They adhere to the following format:

> *X (term being defined) is composed of n parts/types.*

This technique is very useful for beginning paragraphs and sections. It forces organization on the writer and provides expectations for the reader. The following example shows a divisional definition expanding a formal definition:

> *The MacBook Pro is a state-of-the-art laptop computer that has a video display, full QWERTY keyboard, touch pad, and several ports for the interface of various media.*

If this definition were used to begin a section of a manual about the MacBook Pro, most readers would expect the section to continue by discussing, in order, the listed components of the laptop. If the writer did not abide by this forced organization, the readers would understandably and justifiably be confused. So for both writers and readers, divisional definition is a useful tool for organizing the information to be explained.

CLASSIFICATION/PARTITION

"Classification" and "partition" are other useful rhetorical techniques for explaining your subject. They are divisional definitions moved up one level of complexity; they adhere to strict formats as well.

X can be classified as a type of Y.

or

X can be subdivided into the following parts: a, b, . . . n.

These two techniques are used to help organize technical descriptions of objects and processes. The following is an example of a technical description:

The important thing to note in this example is how the writer made use of both classification and partition techniques to organize the information, and to forecast for readers what to expect as they read.

COMPARISON/CONTRAST

Anyone who has ever been subjected to a writing course has been asked to write a comparison-contrast paper, and often the unspoken response has been "Why?" Does such a thing as a comparison-contrast report or paper exist in high-tech industries? The answer is no. But that does not mean comparison-contrast is unimportant. On the contrary, it is a vitally important technique, which appears regularly in proposals and analysis reports. For example, when you write a problem-solving proposal that analyzes several suggested methods for solving the problem, you automatically use comparison and contrast to organize your analysis. This technique may not exist as a pure form for an entire report or paper, but it does help to organize your thoughts for certain types of reports or papers.

As is the case with other writing techniques, the number of ways that you can structure comparison-contrast is limited. It can be organized as a point-by-point analysis (some people refer to this method as alternating), or it can be organized as a subject-by-subject analysis (called "block" by some). The best way to understand the difference is to view the two methods diagrammatically.

Imagine that you want to compare two things (A and B) and that you want to make three statements about each thing. Say you arrange your material as follows: A1, B1, A2, B2, A3, B3, first a statement about thing A, then a statement about thing B, and so on. You have designed an alternating comparison.

3D printing is a generic term used to describe additive manufacturing processes. Unlike computer controlled manufacturing of the past that removed material to produce complex objects, "3D Printing" adds material to a surface or structure over time. This enables much greater object complexity because it is not necessary to fit a tool into the object. Consumer and prosumer 3D printers fall into two general categories: fused filament printing and stereolithography printing. Both technologies process a design into "slices." Imagine slicing an apple into precisely equal pieces. Each slice represents a cross-section of the apple. When all the cross-sections are combined, you have a complete apple. Before printing can begin, a piece of software turns these slices into machine directions.

Fused filament printers work on a principle similar to a hot glue gun attached to a conventional two dimensional printer. A material is extruded at high temperature through a nozzle while the nozzle draws in two dimensions along the slice. The 3D print begins on a platform that moves over time. As each slice of the apple is drawn, the distance between the nozzle and previous slice increases. Over time, all of the slices are drawn and the material cools. This produces an object that looks like an apple.

Stereolithography works on a similar principle but rather than extruding a material with mechanical movement, a laser is used. This laser is focused at a surface of light-sensitive liquid resin. When the laser impulse strikes the resin, the resin hardens near-instantaneously. The laser draws the shape of the slice, just as the filament printer does, and the distance between the platform and the laser is increased. Both methods result in three dimensional objects. Fused filament methods have many more material options, including novel plastics that are primarily wood, metal, carbon fiber, rubber, and more. The flexibility of fused filament printers make them a popular option for models and prototypes. Stereolithography printers have fewer available materials but their simpler design makes for much higher resolution printing. These printers are popular for near-perfect scale reproductions of industrial designs.

Figure 6.1. Technical Description. (Adoniram Sides, Creative Technologist. Used by permission.)

On the other hand, if you arrange your material in this way as A1, A2, A3, B1, B2, B3 (all statements about A followed by all statements about B), you have designed a block comparison. But remember that with block comparisons, you have to provide clear transitions between the blocks so readers can tell how you got from one place to another.

The important question is, when do you use which type of comparison? Block comparison works for all types of comparisons, long or short, simple or complex—as long as you take care of the transitions between blocks. Alternating comparison should be limited to short, simple topics—usually no more than two or three things compared and no more than two or three statements about each one. This criterion is very important. For example, imagine that you are going to compare three things (A, B, and C) and that you will make seven statements about each, and that you are going to use alternating comparison. Off you and the readers go—A1, B1, C1; A2, B2, C2; A3, B3, C3; A4, B4, C4, etc. At about this point, readers start to notice something funny. They begin to see the structure of your information while failing to pay attention to the information itself. No matter how skilled a writing artisan you become, you never want this to happen. Craft never supersedes information in technical writing.

CAUSE AND EFFECT

The rhetorical technique of causal analysis, or determining the causes and effects of events, is the fundamental organizational technique of analytical reports. These reports may analyze problems, or they may explain research and development activities at the conclusion of a project. But at the core of these reports is the treatment of cause and effect.

Causal analysis can be accomplished in any of the following formats.

Single Cause → Single Effect
Example: I dropped the chalk. It broke.

Single Cause → Multiple Effects
Example: I dropped the chalk. It broke and scattered into pieces on the floor.

Multiple Causes → Single Effect
Example: I tripped over the chair and dropped the chalk. The chalk broke.

Multiple Causes → Multiple Effects
Example: I tripped over the chair and dropped the chalk. The chalk broke and scattered into pieces on the floor.

The philosopher in you might argue that all events cannot be reduced beyond the fourth format, multiple causes leading to multiple effects. In reality, this observation is correct. Like the young child listening to the Genesis story, we can always ask, "What happened before that?" But remember: Don't ever confuse technical communication with reality. The two are not the same, no matter how objective you strive to be. As I said at the opening of this book, technical communication is a re-creation of reality based on the needs of an identified audience. A large amount of detail selection goes on in the writing of a report or paper, and our purpose is to cull the irrelevant details so that we can explain our subject. Depending on those purposes and our audience, we might not feel a need to analyze a person's breakfast to determine the causes of a game: design coding error. Nor, perhaps, would we need to report every single event that led to the failure of a lab test, just those that had a significant bearing on the result.

Be careful, however, because this issue brings us to an important aspect of cause and effect: determining what is important and what is not. Although there is a large body of information called formal logic that deals with this subject, we do not have to be concerned with it in detail. Suffice it to say that after we have determined our communication purpose, after we have analyzed our audience, and after we have gathered information, we have to determine what causes and what effects are vital to the needs of our readers. That is the causal analysis to communicate.

DEDUCTION/INDUCTION

Deduction and induction are two organizing techniques that are mirror images of each other. Deduction requires that the writer state a general principle and then give examples to support it. This chapter is a good example of deductive organization. I began with a statement defining clear communication, and I have spent the rest of my time explaining how it can be achieved by examining different examples. Induction is just the opposite. The writer strings together a series of examples or incidents, each edging the reader a little further along. Then comes the punch line, or the statement that ties all the incidents together. Notice that I used the terminology of joke-telling; most jokes are organized inductively. Consider the following story:

> I grew up in a small town in rural North Carolina where on work days people farmed and on off days they either fished or sat around telling lies about farming and fishing. A man we all knew as "Old Pops" was a local fishing legend. While most of us would come back after a day on the lake with 4 or 5 fish, Old Pops never failed to return with less than 30 or 40. Bass, catfish, perch, anything that swam Old Pops could catch. And he did, regularly. When the rest of us told lies about our fishing, Old Pops just sat silent. After all, his actions spoke much louder than our tales put together. Eventually, though, his talents attracted the attention of the game warden. And one day the warden dropped by to suggest that he and Old Pops go fishing together just to see how Old Pops managed

> his success. What could Old Pops say? A week later, he found himself in his boat on the lake with the game warden. As the warden was baiting up, Old Pops pulled a stick of dynamite from inside his shirt, lit it, and tossed it over the side. Seconds later, there was a muffled "whumpf," followed by about a dozen fish floating motionless to the surface of the lake. The game warden, flabbergasted, almost speechless with surprise and indignation, cried, "Old Pops, you can't do that. Don't you know it's against the law, and I'll have to arrest you for it?" Old Pops said nothing. He just reached inside his shirt, pulled out another stick of dynamite, lit it, and tossed it to the game warden's end of the boat. Then he said, "Warden, are we going to sit here and talk all day, or are we going to fish?"

What makes this joke particularly good induction is that the principle is never explicitly stated, even at the end. Rather, it is implied. As a technique for explaining information, however, induction does not have as much potential for use in writing about high technology as does deduction. It is also much harder to control and retain the reader's interest using induction. But it can be used. The following is a more serious example:

> On October 3, at 4:15 a.m., a widespread power outage occurred in the eastern part of the city, where our primary research and development (R&D) facility is located. This resulted in massive losses of data from our division's servers. The specified method for dealing with these sorts of unforeseen disasters is to have multiple backup copies of data stored on external drives. Following the power outage, however, we learned that backups had not been made in approximately two weeks. Consequently, two weeks' worth of work was lost on Project Antelope. Aside from the fact that this will put the project further behind schedule and cost estimates, the lesson that R&D personnel must learn from this is that backups must be made daily.

Notice that in this example inductive organization is combined with multiple causes → multiple effects organization. For most cases, the use of inductive order in writing about high technology should be limited to descriptions that are short and introductory in nature.

CONCLUSION

In this chapter, we have looked at a variety of commonly used rhetorical techniques. You already use these techniques in your everyday conversation, but it is important to be aware of their planned use in writing so that you control them and not the other way around. The time to examine whether you are using these techniques effectively, however, is not when you are writing the first draft of a document. Do it while editing. If you think too much about the technique while you are writing a first draft, you will not be concentrating enough on the information to be communicated. Remember: Form follows function. Or more explicitly: Technique never takes the place of content; it enhances it.

SUGGESTED READINGS

Corbett, Edward P. J. *Classical Rhetoric.* New York: Oxford, 1996.

Kolln, Martha J., and Loretta S. Gray. *Rhetorical Grammar: Grammatical Choices, Rhetorical Effects,* 8th ed. New York: Pearson, 2016.

Williams, Joseph M., and Joseph Bizup. *Style: Lessons in Clarity and Grace,* 12th ed. New York: Pearson, 2016.

CHAPTER 7

How to Use Graphics with Reports and Papers

As graphic design has been made accessible to virtually every writer using Microsoft Word, technical communications have evolved to a point that they contain considerable visual information. Correctly done, graphics (or visuals) are not only informative, but they also draw the reader's attention to information the writer chooses to highlight. They can carry much more information per space in a document than the same amount of text can. And if one definite trend can be counted on to continue in writing about high-tech subjects, it is an increasing reliance on visual communication. Visual communication aids all readers, but it is particularly useful in international situations because less translation is necessary.

In paper-based documents, two kinds of graphic communication are used—tables and figures. They differ in format and in the way they are incorporated into the text of a report or paper. For example, formal tables are numbered and have a title; all of this information is traditionally placed *above* the table. Figures also have numbers and titles (sometimes called captions), but the information is traditionally placed *below* the figure. The only difficulty is remembering what is a figure and what is a table. Actually, it's simple: everything that is not a table is a figure. Furthermore, there are only two types of tables—informal and formal. And only one of those, the formal table, has identifying numbers and titles. In this chapter, the most common forms of graphics found in writing about technology will be examined and explained. Each has its best use; writers must determine the nature of the information that is to be communicated and choose the appropriate graphic accordingly from among the many possible options that Word provides.

INFORMAL TABLES

Informal tables are simply lists. As such, they are not numbered in a sequence throughout a report or paper. Rather, they are an extension of the text. But remember, they should be physically separated from the text they accompany by a sufficient amount of white space before and after the list and by additional margin space to the left and to the right of the list. This physical separation calls attention to the information in the list, and it is what makes an informal table a visual technique in the report or paper's design.

This book has several informal tables in it. Figure 7.1 is an example. Remember that the figure number and title are for the graphic in *this* text. The original had neither.

FORMAL TABLES

Formal tables require more formatting than lists. They are usually separated from the text by top and bottom rules (lines) and by a table number and table title, both of which are placed *above* the table. Formal tables should have clear column and line headings, both of which should accurately describe the types of information found in the table. If necessary, both column and line headings may have subheadings. The individual cells for data, or information, should be spaced adequately so they are not crowded and hard to read. If an explanatory note is needed for any information contained in the table, it is placed immediately below the table as a footnote. If the entire table was copied from another source, the title of the table receives a reference—either with a superscripted Arabic numeral or a numeral in parentheses, depending upon the documentation style the writer is following. Figure 7.2 is an example of a formal table. (Again, the figure caption and title are only for use in this book.)

Currently, the four most popular laptops, by market share in the United States, are:

- HP
- Lenovo
- Dell
- Apple

Figure 7.1. Informal Table.

Table: Top Notebook Brands Worldwide by Shipments, 2014~2016

Ranking	2014 Company	Market Share	2015 Company	Market Share	2016 Company	Market Share (E)
1	HP	20.1%	HP	20.5%	HP	20.7%
2	Lenovo	17.5%	Lenovo	19.9%	Lenovo	20.0%
3	Dell	12.3%	Dell	13.7%	Dell	14.0%
4	ASUS	11.0%	Apple	10.34%	ASUS	10.7%
5	acer	10.0%	ASUS	10.31%	Apple	10.3%
6	Apple	9.3%	acer	8.9%	acer	9.0%
7	Toshiba	6.6%	Toshiba	4.2%	Samsung	2.4%
8	Samsung	2.7%	Samsung	1.7%	Toshiba	1.6%
9	Vaio (Sony)	0.6%				-
	Others	9.9%	Others	10.3%	Others	11.4%
Shipment Total (Unit: M)		175.5		164.4		159.2

Note: Vaio (Sony) was included in "Others" category in 2015.

Figure 7.2. Formal Table. (Trendforce. "Trendforce Reports Notebook Shipments Totaled 164.4 Million Units in 2015 with Apple Gaining Greater Market Share Annually." February 16, 2016. Available online at http://press.trendforce.com/press/20160216-2327.html)

LINE GRAPHS

Line graphs are used to show changes in the state of something over a period of time. For them to communicate effectively, the graph axes must be labeled clearly and descriptively, with units of measure also marked. In multiple-line graphs, the lines should be differentiated with symbols that are explained in a key or by color, if full-color printing is used, rather than with various dotted and dashed lines. This use of symbols will enable you to use the graph more effectively as a slide in a presentation, should that be required. Figure 7.3 is an example of a multiple-line graph.

BAR GRAPHS

Bar graphs are used to compare the size or number of items. Bar graphs may be designed on axes similar to line graphs, but neither axis will signify a change in either state or time. Once again, the axes must be clearly labeled. Any units of measurement are also marked, and the actual measurement is usually listed at the top of each bar. Although many graphic design functions, including those in Word, provide for the ability to create three-dimensional bars in a bar graph, you are better off not using these, since it can be confusing as to where the top of the bar is, since it is a three-dimensional image superimposed on a two-dimensional background. And no one wants to get involved with creating an appropriate z-axis to indicate three dimensions. Figure 7.4 is an example of a bar graph.

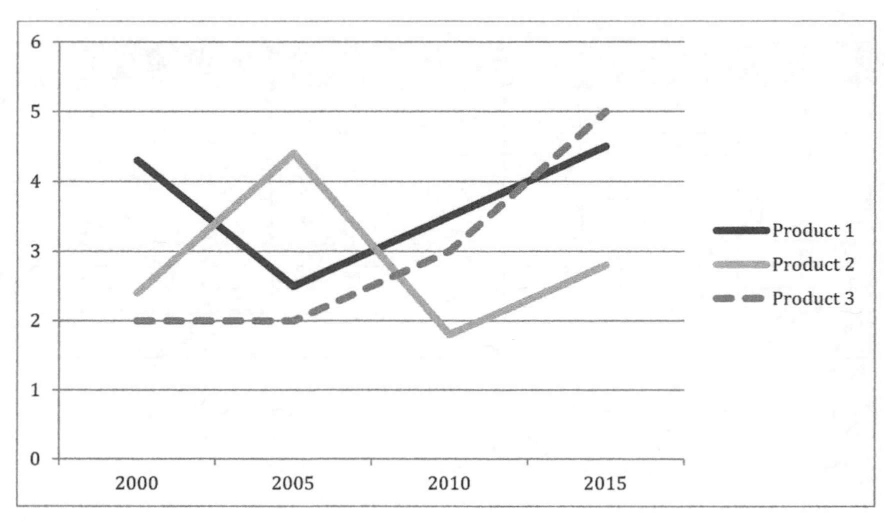

Figure 7.3. Multiple-Line Graph.

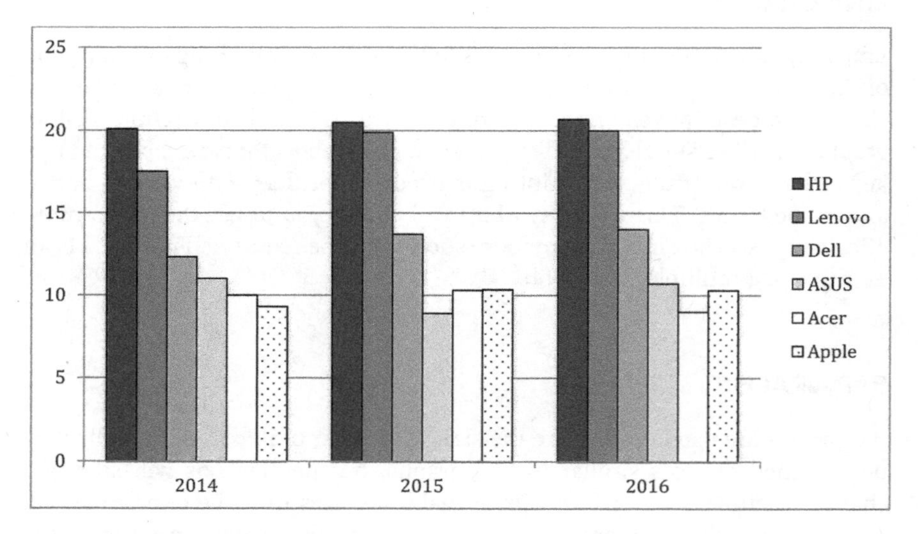

Figure 7.4. Bar Graph.

DIVIDED BAR GRAPHS

Divided bar graphs are similar in appearance to bar graphs, but divided bar graphs compare percentages of a whole rather than relative size. Bar labels and percentages are often placed within the bar when space allows it. If there

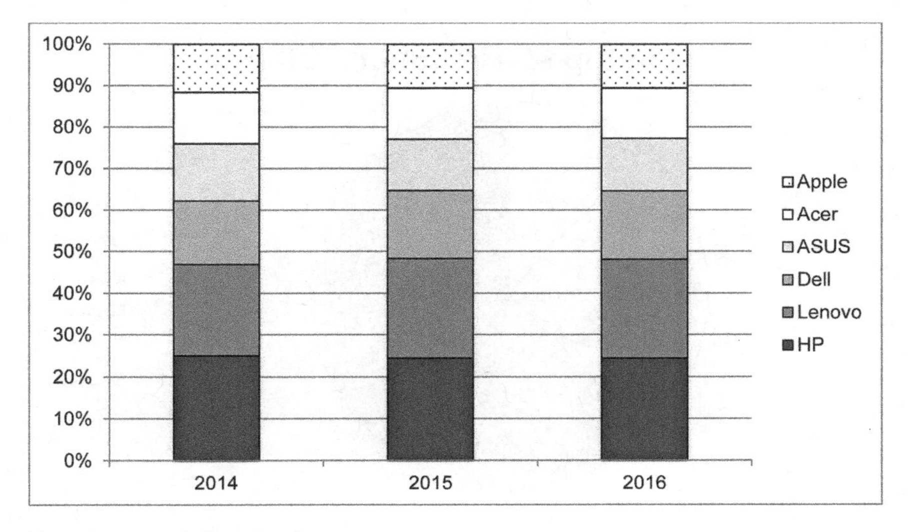

Figure 7.5. Divided Bar Graph.

is not enough space, this information is placed outside the bars, and faint lines are drawn to connect the information to the appropriate bar. Figure 7.5 is an example of a divided bar graph.

CIRCLE GRAPHS (PIE CHARTS)

A circle graph, or pie chart, serves the same purpose as the divided bar graph: it compares percentages of a whole. It is probably more effective visually than a divided bar graph for this purpose because readers can get a clearer sense of the whole. Circle graphs are complete; bar graphs look as if they could be extended. With the circle graph, labels and percentages can be placed inside each wedge if space allows. Often it does not, and even when it does, labeling is made difficult by the odd angles of the wedges. For these reasons, labels and percentages are usually placed outside the circle graph and connected to the appropriate wedge with a line. Remember, also, that the largest wedge begins at 12 o'clock and moves clockwise. Figure 7.6 is an example of a circle graph.

PICTOGRAPHS

Pictographs used to be restricted to the domain of graphic artists, but 21st-century word processing programs enable less-than-artistic writers to use

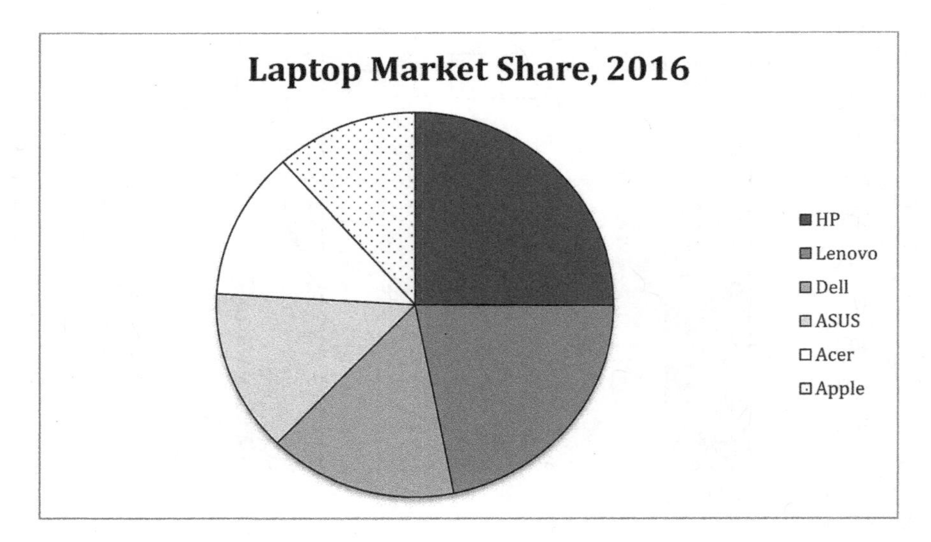

Figure 7.6. Circle Graph.

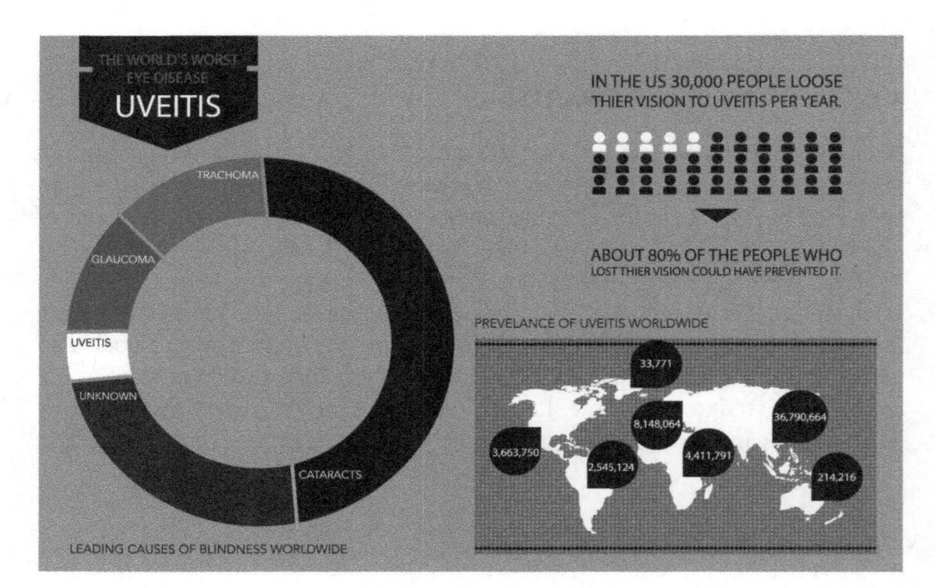

Figure 7.7. Pictograph. (Designed by Danielle Foster, www.designsbydani.net. Used by permission.)

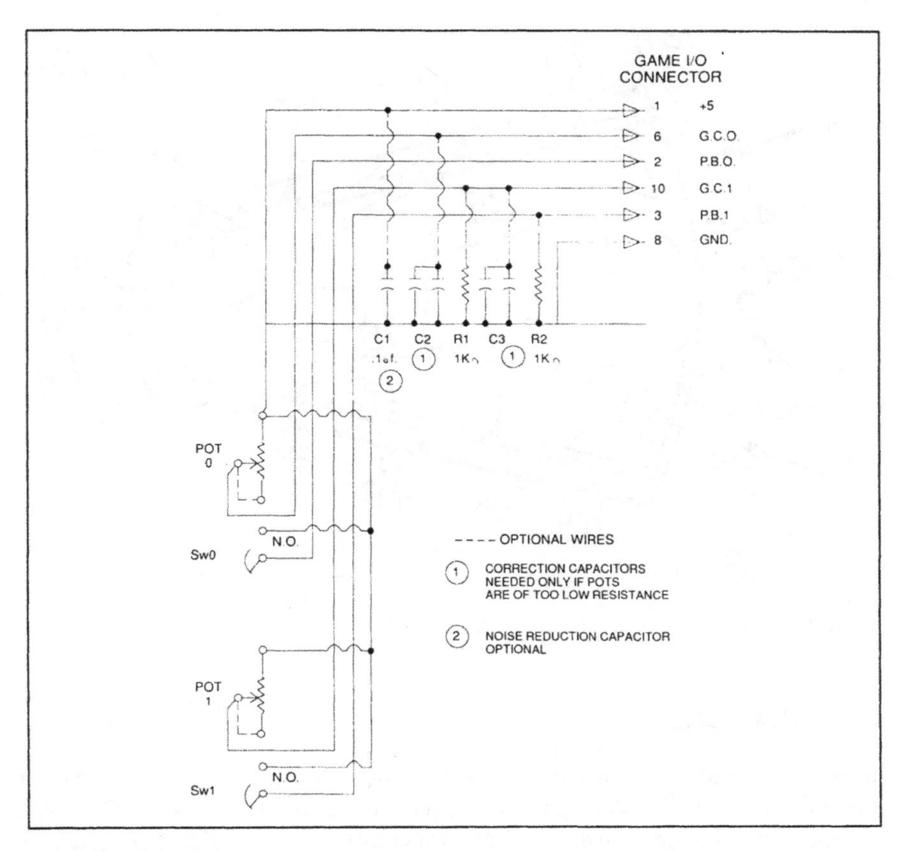

Figure 7.8. Schematic.

these powerful visual displays. They are more illustrative in their design than ordinary graphs and can be particularly effective in attracting the attention to information you wish to communicate visually. Figure 7.7 is an example of a pictograph.

SCHEMATICS

Schematics are particularly important to writers of reports and papers in the high-tech industries. They are used to show the relationship of parts, a type of visual outline. For schematics to work, however, they must be very clear and absolutely uncrowded. Labels and headings should be placed where they can be read. Circuit drawings and flowcharts are familiar schematics. Figure 7.8 is an example of a schematic.

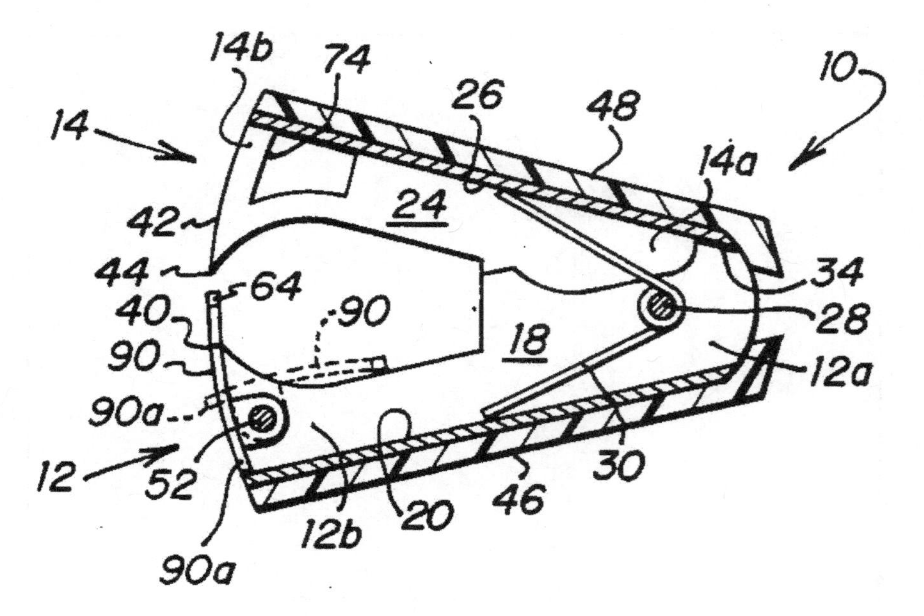

Figure 7.9. Illustration. (U.S. Patent 4674727, June 9, 1986.)

ILLUSTRATIONS

Illustrations are the final type of graphic to be considered. They are two- or three-dimensional renderings of an object. To create these successfully, you should have considerable experience with computer illustration tools (such as those provided in Adobe Creative Suite), as well as the graphic design capability to use those tools well. Otherwise, leave these up to the company graphic artist or to a freelance artist. If you draw them yourself, be sure to orient the illustration in a way that makes sense to the reader. Also, label all important parts clearly and give appropriate dimensions. It is important that readers, particularly unsophisticated readers, are able to tell whether what is being illustrated is the size of a nuclear reactor containment vessel or an element of nanotechnology. Figure 7.9 is an example of an illustration.

THE SPECIAL CASE OF ELECTRONIC MEDIA

More and more documents now exist in online environments. Several factors have contributed to this trend: virtual storage requires no physical space, as do hard copy documents; environmental concerns (less use of paper); costs (online documents greatly reduce publication costs, and if many users still prefer print documents—and they do—they can print the documents themselves, thereby passing on a printing cost to the customer).

But using visuals in an online environment is categorically different from using them in paper documents. Consider the following rules of thumb:

- Keep in mind that one page equals 3–6 screens or windows.
- Use "golden rectangles"—shapes that are 1.6 times as wide as they are tall—because they equate with the acuity of human vision, an ellipse that is 1.65 times wide as it is tall.
- Remember that colors are displayed differently on screens than on pages; screens produce color by adding various amounts of the primary colors for light (red, green, blue); printing creates color by subtracting various amounts of cyan, yellow, and magenta, while adding black as a fourth color. Red text is never a good idea in online environments because it strains the reader's eyes, appears fuzzy, and even almost appears to be moving.
- Use tabular formats, such as lists and boxes or tables.
- Keep lines 12–14 characters long.
- Standardize window placement and organization.
- Present related information vertically, rather than horizontally.
- Use labels or headings liberally.
- Despite the fact that monitor displays are now high definition, printed pages are still much higher definition, so be certain that fonts and other typographical techniques work well in the online display environment. A good rule of thumb is that if you can't read it, your customer/reader can't either.

PLACEMENT OF GRAPHICS

One point that has not been mentioned so far is the placement of graphics or visuals. If the graphic or visual helps explain an important point to the readers, place it as close to that point as possible. Usually this means on the same page. If the graphic is included for reasons of documentation accuracy (to cover yourself), it is acceptable to place it at the end of the report in an appendix. For articles and papers submitted to magazines and journals, note the position of the graphic (by name and number) in the manuscript. Then place the graphic at the end of the manuscript. The editors will see that it is properly placed.

CONCLUSION

In this chapter, a variety of graphics that writers can use with reports and papers have been examined. When chosen and rendered accurately, with your

Rhetorical Purpose	Type of Graphic
Describe Organization	Pictograph, Flowchart
Compare/Contrast	Pictograph, Pie Chart, Bar Graph
Describe Parts	Schematic
Classify	Table, List, Pictograph
Describe Change of State	Line Graph, Bar Graph
Relate Date to Constants	Line Graph
Describe Process	Pictograph, Flowchart
Describe Steps in a Decision	Flowchart, Pictograph
Describe Multiple Responsibilities	Flowchart
Describe Proportions	Pie Chart, Bar Graph
Describe Relationships	Table, Line Graph
Describe Causation	Flowchart, Pictograph
Describe Entire Object	Schematic, Illustration, Photograph
Present Raw Data	Table, List

Figure 7.10. Graphic Selection Chart.

purpose and audience in mind, they can be vital additions to written communication. Be sure that every graphic you use is clearly explained in the text. This chapter ends with a figure (Figure 7.10) that will help you determine which graphic is most appropriate.

SUGGESTED READINGS

Cook, Gareth, and Robert Krulwich, eds. *The Best American Infographics 2016.* New York: Mariner Books, 2016.

Foreman, John W. *Data Smart: Using Data Science to Transform Information into Insight.* Hoboken, NJ: Wiley, 2016.

Knaflic, Cole Nussbaumer. *Storytelling with Data: A Data Visualization Guide for Business Professionals.* Hoboken, NJ: Wiley, 2015.

Lankow, Jason, John Ritchie, and Ross Crooks. *Infographics: The Power of Visual Storytelling.* Hoboken, NJ: Wiley, 2013.

CHAPTER 8

How to Use Writing Tools

Writing tools are ubiquitous in the 21st-century workplace, so much so that virtually all of us never give them a single thought. We write on desktops, laptops, and tablets; we text on phones; we are engaged in an almost 24/7 immersion in social media that we actually do at times apply to our professions, instead of simply posting political rants and cat videos.

Despite the fact that those of us who adopted these technologies during our careers rarely think critically about them, we should. And the generation that has never been without these tools—you definitely should.

WORD PROCESSING

First, a little history. Word processing is where the writing technology boom began. And it is omnipresent. We outline, draft, polish, and revise our work without ever committing print to paper. Then, if necessary (and increasingly it is not), a laser printer can make our words look beautiful, in a variety of fonts that were available only to professional printers throughout much of the last 500+ years. But "look" is a key word here. Lightning-quick technology does not absolve any of us from the responsibility of paying attention to the details described elsewhere in this book. *There is plenty of research into the practice of writing that suggests we actually spend less time—now that banging out words with a keyboard is so fast—considering the quality of what we have written, all the while producing massive quantities of it.*

We hammer out document after document, pay scant attention to what we have typed, and blast it off to readers. In fact, in the public relations discipline, there is even a term for it—"e-mail blast." But we should not forget the principle of getting our communication correct before we commit it to the readers.

Many people would find it helpful to write a draft very quickly, edit it, and then revise it. That works for me, and a lot of professional writers agree. As mentioned previously, try to avoid the temptation to write and edit at the same time. Most of us are not good at those sorts of mental gymnastics.

One final word about word processing, especially in professional environments: Be prepared for disaster. If Murphy's Law applies anywhere, it applies here. But don't let it apply to you. Back up copies of important work on portable media so that if your computer decides to feed upon one, you will have others safe for use later. Print out hard copies from time to time. Yep—I actually said that. And it does not consign me to the Paleolithic Age. Two years ago, the happened to me: I was invited to give a keynote address to a conference in Colorado. I wrote out the script for my presentation. I saved it to a USB drive, and placed that in my checked luggage at the airport. I e-mailed it to myself, so I could access it on my iPad. I printed and took a hard copy in my briefcase as my carry-on luggage. When I arrived at my hotel in Colorado, the USB drive was missing. (I googled this and learned that, in fact, the Transportation Security Administration [TSA] does remove these things from checked luggage on occasion.) I went to the convention center ahead of time to check out the arrangements for presenters (we'll cover this in a later chapter), and learned that my iPad would not communicate with the presentation hardware and software—no Bluetooth arrangements. So, I took my hard copy to the nearest copy center and printed the PowerPoint slides so I could use them as physical handouts during the presentation. These things happen, still; don't let them happen to you. Have a Plan B, and even a Plan C.

SPELL CHECKERS AND GRAMMAR CHECKERS

These are valuable tools that are in play when we write using such programs as Microsoft Word. But they are not infallible. Spell checkers now have much larger vocabularies than virtually any native English speaker, but there are things they cannot do. For example, if you write: "The book was read," but you mean: "The book was red," Word spell checker thinks each of these is perfect.

So, check the spell checker, and always read over your work for errors before you consign it to the professional record of your employer, as well as before you send it to readers, customers, etc.

Grammar checkers are even more unreliable. For one thing, grammar is considerably more complex than spelling. As a result, it is more difficult to design a software program that does a thorough job of checking a writer's grammar. Another problem is that there is less widespread agreement about what constitutes correct grammar. Turn a grammar checker loose on a famous author's work and watch it locate error after error. Acceptable standards for grammatical usage change more readily than spelling standards. For example, is it "None are . . ." or "None is . . ."? My Microsoft grammar checker correctly points out that the first one is incorrect, but the clear majority of native speakers now say that the second one is. That is how English evolves, and the second one is now acceptable except in the most formal of communication circumstances.

CONCLUSION

In this chapter, we have taken a quick look at the major tools of the writing trade. They are continually improving, making the task of sorting out what's difficult for you ever easier.

SUGGESTED READINGS

University of Chicago Press. *The Chicago Manual of Style: The Essential Guide for Writers, Editors, and Publishers,* 16th ed. Chicago: University of Chicago Press, 2010.

PART III

How to Write a Paper or Report

CHAPTER 9

How to Organize a Paper

Professionals in science and technology often have the opportunity or the desire to publish outside their own particular organizations. They have many reasons for doing this: professional prestige (peer-reviewed journals), money (provided the magazine, newspaper, blog or web site pays for the contributions), professional advancement, or because someone made them do it. Whatever the reason for submitting an article to the media outlets listed above or a paper to a professional journal, the challenges and the rewards are categorically different from those encountered in writing internal reports within your place of employment.

The first problem that confronts the author is defining just what is being written. Is it a paper, article, or report? While some confusion exists over the differences among these three types of writing, the typical distinction between papers and articles is that a technical or scientific paper is formal, intended for a relatively limited audience, and published in a professional journal (such as *The Journal of Pediatrics, Nature,* or any of hundreds of others); an article, on the other hand, is informal, intended for a wide audience, and published in trade journals or the science and technology sections of newspapers, or in blogs or web pages. Although papers and articles might communicate the same information as reports, reports are generally classified into various types of formal documents: proposals, feasibility studies, interim reports, and so on. There is, however, enough overlap in the forms of communication that the following chapters on writing papers introduce topics common to reports, papers, and articles.

AUDIENCE

The first problem in writing papers is the audience. It is large, anonymous, and very difficult to analyze. Trying to write for such an audience is akin to writing in a closet with the door closed. You don't know what's on the outside. Even if you are comfortable with the audience-analysis system presented in this book, you quickly realize that it does not provide as much information in this environment as it does for internal reports. Answers to the audience-analysis questions are automatically vague. You can't identify readers by name and position, and even if you obtained a copy of the circulation list for a journal (not a bad idea), it would only do you a little good.

The audience issue for articles and papers is bleak, but it is not hopeless. The best way to learn about your prospective audience is to contact the journal, magazine, or newspaper editor and ask; the same strategy works for blogs and websites, too. Each publishing outlet available to professionals in scientific and technological disciplines is intentionally aimed at a specific market; in fact, with scientific and technological fields being what they are, each market has many outlets. Let the editors of these do some of your audience analysis for you. In addition, be sure to ask for a sample copy of the publication or a link to the blog or website so that you can use it to analyze the audience.

ORGANIZATION AND STYLE

Once you have decided who the audience is, analyzed papers or articles or postings that have been published in your target publication, and determined what information would interest your readers, you must solve the problems of organization and style. You will have to fit your article or paper into the mode expected by readers of the publication. It's the only way. You won't be able to change editorial policy, so to be published, you will have to play the game by their rules. Contact editors and web-content managers to ask for their submission requirements and style guides.

Once again, examine successfully published papers, articles, and posts. Look at things such as sentence structure, sentence length, use of personal pronouns, use of the passive voice, overall writing style, illustrations, photographs, charts, tables, use of headings within the paper—anything that might affect the way you present your information. Remember also the two ways of perceiving—intuition and sensing. Use headings and orientations liberally throughout your text so intuitive readers can skim. Provide plenty of details within paragraphs and sections so that sensing readers can follow your train of thought. Also, find out about the accepted referencing styles for professional journals. The editors and content managers can answer all these questions, allowing you to spend your time more profitably by gathering information and organizing it. The more you know about what is expected of a submitted paper, article, or post, the more likely you will be to get an acceptance.

WRITING THE ARTICLE OR PAPER

Based on what you find out by examining the publication, you will be faced with a different sort of communication problem with different solutions. Writing a paper or an article about science and technology is not the same as writing a report, even if the topic is the same. Different skills and tasks are required, all of which must be mastered by you to be a successfully published writer. The next three chapters are devoted to the most difficult tasks of writing articles and papers—writing the discussion, the conclusion (or exit), and the lead. You might wonder why they are in this order. Why not start at the beginning?

In fact, you are. For a writer, the beginning is the middle. You cannot clearly lead readers to an understanding of your subject if you do not first know where you are going yourself. Beginning in the middle is time- and task-effective. It allows you to write as soon as you have your information, without wandering off on tangents and into blind alleys. After you have written the middle and the end, the hardest part—writing the beginning—will be much easier.

You might also wonder why this chapter focuses mainly on writing articles. One reason is that the intended readers of this book are more likely to write articles than peer-reviewed papers. Another and more important reason is that the format for articles is much more flexible, more open to the design whims of individual authors. Those of you who are faced with writing a paper will find formats rigidly prescribed by the target journals. A quick examination of the journal will also show how the individual sections of a paper are to be written.

Figure 9.1 depicts a general approach for organizing articles about science and technology intended for trade journals, magazines, newspapers, websites, and blogs.

CONCLUSION

One final thing: rejections. Expect them. They are a reality in the lives of all professional writers. They are an affront to your integrity, or your abilities. Most often they simply mean that what you have to say does not fit with the editor's idea of who the audience is, or what they are interested in. Look for

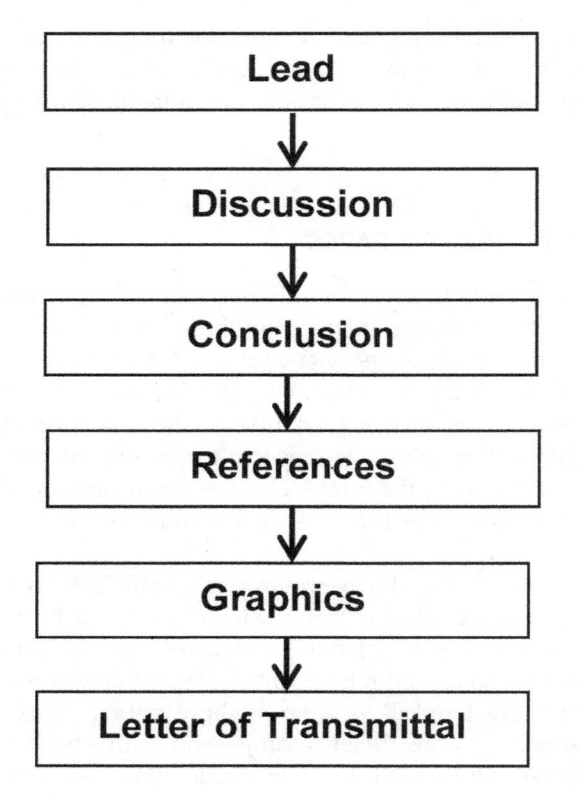

Figure 9.1. VOS for Articles.

another publication and try again. If you are willing to make the effort, write a follow-up to the rejection, asking for more specific information as to why it was rejected, and asking whether you may resubmit it provided necessary changes are made. This tactic does not work all the time, or even perhaps most of the time, but if it works once, it is worth the effort.

Suggested Readings for chapters 9–12 appear at the end of chapter 12.

CHAPTER 10

How to Write the Discussion or Body of an Article

Before we can consider how to write the discussion or body of an article, we should consider what makes this part of an article work. Once again, the answer is organization. The body of an article is what supplies readers with the specific information they might be seeking. It is the place for facts and figures, nuts and bolts; in a word, details. Just as it will be important to tell readers what you are going to do before you do it, it is equally important to do it after you have told them you would. This back-and-forth is how a well-designed lead and body of an article should work together. The body must present the information forecasted in the lead, and it must present it so that readers can understand and use the information. Otherwise, you are wasting a busy reader's time. That is why it is safer to write the body first; then you know what to include in the lead.

The basis of all well-written discussions is order. We will consider three common orders in this chapter: (1) reader-centered order, (2) writer-centered order, and (3) subject-centered order. Notice that each of these orders emphasizes one of the three parts of all communications—audience, writer, and message.

READER-CENTERED ORDER

A reader-centered discussion or body of an article seeks, as its main goal, to satisfy the needs of the reader. But, you say (especially after having read the chapter on audience analysis), isn't that the goal of all communication? Yes, it is; but the reader-centered discussion puts more emphasis on reader needs than do the other discussion orders. The following outline depicts a reader-centered order for the body of an article.

Outline of a Reader-Centered Discussion

1. The first section of a reader-centered discussion presents specific information about the topic of the article. It defines the problem or what is at issue in the article and provides a brief statement of its resolution. It can then suggest to the reader what to expect in the rest of the article if this has not been done already in the lead. If the article and the lead are quite long, the first section of the discussion might repeat (briefly) a forecast of the remaining structure of the article, even though it was initially mentioned in the lead. As we saw in the section covering purpose statements in Chapter 4, the lead plus the first section of a paper's discussion constitute a purpose statement introduction.
2. The second section of a reader-centered discussion presents criteria, or assumptions, that will be used to evaluate the validity of the writer's conclusions. This is a logical extension of the first section's resolution of the central issue of the article. Establishing these criteria erects a frame around your work, limiting what your readers can take issue with. Such limits also make your job of organizing the information and the reader's job of understanding it easier.
3. The third section of a reader-centered discussion (usually this contains several subsections) provides the support for your conclusions. This information can be organized in two ways: from most important point to least important, or from least important point to most important. Each way has advantages and disadvantages. From most important to least important is easier for the reader to understand. You have told them what they most need to know early in the article, tying it in closely with the lead and the first part of the discussion. But if you use this arrangement, the reader's interest might lag toward the end of the discussion as your material gets less and less important. From least important point to most important is more dramatic and suspenseful; the reader's interest tends to build throughout the article. But most articles about science and technology do not have drama and suspense as two of their primary aims, although—especially with science writing—there are notable exceptions. More importantly, however, this arrangement makes it difficult for readers to follow the writer's train of thought. It is easy for readers to become confused. So, what is the answer? In most cases, the best arrangement for articles about technical or scientific subjects is from most important point to least important. Just remember that you should not belabor your points toward the end of an article to the extent of boring your readers. As your material becomes less interesting, your writing pace (and the reader's reading pace) should quicken. Be sure to look at Chapters 19 and 20 to see how to do this.

4. The fourth and final section of a reader-centered discussion restates the most important point the writer is trying to make, perhaps showing how the point has been supported in the article. This section should lead quickly and smoothly into the exit of the article.

The following example presents excerpts from a reader-oriented discussion. The entire discussion was too long to reproduce here, but the four parts of it are nonetheless identifiable in these excerpts:

Problem Definition

The mealing condition evident on Printed Wiring Assemblies can be directly related to the card and system exposure of a humidity test, as Printed Wire Assemblies in systems not exposed to humidity do not exhibit mealing. Additionally, all the required contractual testing has shown that the functional performance of GTE products is not affected by mealing on assemblies. With the functional performance of the product unaffected and the mealing being relatively discrete, even though it covers a large area of the assemblies, the consensus was that GTE does have mealing but it is of a minor nature and that the long-term reliability of the assemblies will be unchanged.

The plans of the task team were therefore directed down two paths: first, to verify the integrity of the process to clean flux (ionic contaminants) that could affect reliability from the assemblies; and second, to perform investigations and testing to determine what is causing the mealing.

Criteria

Specification requirements for mealing are found in MIL-P-28809, which is called out by MIL-P-11268 and MIL-STD-454.

Mealing is identified as a major defect in MIL-P-28809. However, as will be shown later, it is called major under the assumption that ionic contamination is the cause.

Support

The most important investigation was determining the level of ionic contamination by verifying the effectiveness and accuracy of the Omega meter.

Specification MIL-P-28809 requires a resistivity reading of 2.0 Megaohms/ cm or greater. Testing samples per MIL-P-28809, Paragraph

4.8.3., and in the Omega meter, average readings of 11.0 Megaohms/cm and 10.1 Megaohms/cm respectively were obtained. This is a negligible difference and well above the specification requirement.

Other testing conducted up to this date has shown mealing can be created on bare boards that were cleaned with processes other than aqueous and on boards that were not exposed to flux and wave solder. This result supports the assumption that assemblies are being properly cleaned and directs testing away from the solder/cleaning process.

Summary of Main Points

From investigations to date, the government has established that the cleaning processes are effectively removing ionic contamination from the Printed Wire Assemblies. In addition, GTE's experience from in-house and field testing has exhibited no reliability or functional performance degradation due to the presence of discrete Printed Wire Assemblies.

WRITER-CENTERED ORDER

A writer-centered discussion or body of an article seeks to portray the process the writer or writers followed during the research and/or development of the article. For the most part, this order is restricted to subjects that are procedural in nature, and that can be structured as a narrative. When the subject lends itself to this order, or when the writer's purpose suggests this order, it can be a very effective way of organizing the body of an article. It draws upon the direct experience of the writer with the subject, and often tells that experience in the first person. For an excellent example of this structure, look at Siddhartha Mukherjee's *New York Times* Best Seller, *The Gene: An Intimate History,* a popular, nonfiction account of genetic research. The following outline shows how writers might organize information in a writer-centered discussion.

Outline of a Writer-Centered Discussion

1. The first section of a writer-centered discussion presents how the writer discovered the topic. Depending upon the topic, this section might be on the need for a new software product or a section on the writer's day-to-day activities in a field of science or whatever. This section extends the lead and begins the narrative.
2. The second section of a writer-centered discussion develops the narrative by describing the writer's experiences with the topic. This section is usually the longest part of this type of article. For it to be successful, it generally follows a chronological order and leads to the final section.

3. The final section of a writer-centered discussion wraps up the narrative, usually giving the readers some insight and understanding, a resolution of sorts, of what the writer has been through. This section should lead smoothly to the exit of the article.

Writer-centered articles are often entertaining as well as informative. As such, we are more likely to find them in the popular press and blogs than in peer-reviewed or trade journals. A consistent (and excellent) exception to that is the trade journal *Design News*. You will find writer-centered narratives from engineers and designers in that journal on a regular basis. One last point about writer-centered articles: The market for them is highly competitive, and the success rate for publishing them is fairly low. But they are enjoyable to write and are a break from the type of writing most of us find ourselves doing on a regular basis.

SUBJECT-CENTERED ORDER

This order is the one most commonly used for the bodies of articles or for the discussion sections of reports. It is serviceable and easy to use for any subject. Because it is used so frequently, this order will not make the subject stand out. Writers, however, can generate emphasis through a variety of other stylistic devices, which will be considered in a later chapter.

While this explanation may sound as if I am downplaying the importance of subject order, it should not. You will find yourself using this order more than the other two combined. It is that flexible, and it works that well. Another advantage of the subject-centered order is that it can be combined with the writer-centered order. If your topic is broken down into subtopics and one or more of those subtopics is procedural in nature, you might find such a combination advantageous to your article. The following outline depicts the general-to-specific organization of subject-centered order:

Outline of a Subject-Centered Discussion

1. The first section of a subject-centered discussion presents the first subsection of the topic. The topic should have been clearly subdivided in the lead so that the readers can anticipate the order of the discussion, and the discussion should treat the subtopics in the order they were listed. A common mistake is for writers to neglect to follow their own organization, resulting in the complete befuddlement of the readers. In this type of discussion structure, too, the writer can arrange subtopics from most important point to least important and vice versa.
2. The second section of a subject-centered discussion, and every section thereafter, presents the remaining subtopics in the order in which they

were listed in the lead. The final section of this type of discussion should lead the reader to the conclusion, which is almost always a summary.

The following selection, written for trade journal readers in the tourism field, is an example of a subject-centered discussion. The article describes goals for a major re-envisioning of tourism for a small, rural district in a northeastern state. Notice how the introduction sets up the discussion by forecasting the information that will follow. The discussion which resulted was simply a catalog of the forecast information, with individual paragraphs devoted to each topic:

> The focus of the Quaboag Hills Region Tourism Initiative is a collaborative project between the Quaboag Valley Chamber of Commerce (QVCC) and the Quaboag Valley Community Development Commission (QVCDC). The campaign to promote tourism in the region will focus on generating awareness of the region among residents of the northeast United States. This will in turn increase the number of visitors to the region and stimulate economic growth among businesses related to tourism activities. In addition to raising awareness and stimulating growth, we have identified the following goals:
>
> - Create a thorough, updated catalog of outdoor recreational activities
> - Develop a methodology for regular updates of this information
> - Create a combined map & brochure that shows where and when activities take place
> - Distribute maps and brochures to visitor's centers throughout the Northeast
> - Coordinate overall promotion of activities of interest to tourists.
>
> (unpublished)

CONCLUSION

In this chapter, we have looked at common ways used to organize the bodies of articles. Remember that even though your audience might share your considerable expertise on the subject, flawlessly clear organization will always lessen the reader's work and enhance their appreciation of you.

Suggested Readings for chapters 9–12 appear at the end of chapter 12.

CHAPTER 11

How to Write the Exit

Writing a good exit, or conclusion, is the second hardest task in writing an article. (See chapter 12 for the hardest task—writing the lead.) And unfortunately, there are no formulas that guarantee successful exits all the time. But good exits are vital to your communication, so time must be spent developing techniques that work for you. If the axiom that practice is the only road to success is true anywhere in the field of writing, it is in the writing of exits. The more you practice, the more you will develop what has been called a "felt sense" as to when they are right. In other words, you will just know intuitively when an exit is right, when it works for you and for the readers. This advice may sound like typical writing-teacher hocus-pocus, but it isn't. Trust me.

Or if you don't regularly trust what you are told about writing and want more specific guidance, all is not lost. Even though formulas for successful exits don't exist, it isn't as bleak as that may make it seem. Exits tend to follow an order of their own, and we can group them as follows in this chapter.

SUMMARY EXITS

Summaries are the most common type of exits written for articles in scientific and technological fields. A summary exit of a reader-centered article might reintroduce the central issue of the article and restate the solution more fully than it was initially forecast in the opening section of the discussion. In a writer-centered article, it might restate the objectives of the writer's work or the problems or needs that caused the writer to do the work in the first place. In a subject-centered article, it might simply restate the main points made in the paper.

One potential problem with using summary exits is the tendency to confuse them with a part of a report that is also called a summary. They are not the same. A report summary is a distinct section intended for a distinct audience. If you think of it as an executive summary and realize that it is placed at the beginning of a report, the confusion will be avoided. The purpose statement introduction discussed in chapter 4 can easily be expanded into an executive summary.

Another source of confusion, even if we limit our discussion to articles and papers, is that many people call an abstract a summary and vice versa. An abstract also differs from a summary exit; it is a synopsis of an entire report or paper, placed before the beginning of the paper and intended to be used independently of the paper. For example, imagine that you are doing some literature research that requires examining many papers. The abstract/summary, if well written, will enable you to determine whether you have to read the entire paper because it will contain the important aspects of the paper.

Following is an example of a summary exit to a writer-centered report. It is an important example because it shows that writing about computer technology often occurs in unexpected places outside high-tech industry, in this case an elementary school:

> Conclusion
> Because of the influence of the computer technology in all disciplines, our school board has directed us to build, equip, and manage a computer learning web design center at Pineville School for the purpose of teaching web design and development to high school students. The school board further directed us to develop a computer supported curriculum for all course subjects. To accomplish this goal, I was instructed to write a proposal detailing the steps needed for such a project.
>
> I began by studying all the unused rooms in the building, their location, size, and general condition. The custodian and I inventoried and inspected every piece of surplus furniture in storage. I have spent considerable time with Best Buy salespeople determining the following: what hardware is available that is compatible with what the school already owns, what price Best Buy will charge for various lines of equipment, and what extended service plans are available. In addition, we have explored various web design products from popular "drag-and-drop" software to html coding.
>
> Based on my study, I conclude that it is reasonable and practical for us to build a computer learning web design center to teach html coding at a cost that is within our funding limits.
>
> Using three summer workshops and in-service programs, we can do the following: train inexperienced teachers in the basics of coding,

develop a curriculum to teach the students the basic concepts of html, buy software to supplement teaching, and establish a committee to develop long-term goals.

LOGICAL CONCLUSIONS

In a persuasive or argumentative article, the writer will most likely want to lead readers to a logical conclusion. At the end of the article, the readers will reach the same logical conclusion if the writer organized the topic right. Although each type of discussion organization can be used for this type of conclusion, it is most commonly found with reader-centered discussions.

APPLICATION OF BASIC POINTS

The exit to an article might directly show the readers how they can use the information presented in their own companies and their own jobs. This type of exit, too, is most often found with reader-centered discussions, but it will work equally well with subject-centered discussions.

FORECAST OF FUTURE EVENTS

In trade journals, this ending is probably the most important and most desired type of exit. It provides readers, particularly those in management and design functions within technological or scientific organizations, with information that they most need to know—where the industry is going in the next few months or years. This type of exit can be applied to each of the discussion orders we considered. It works very well with reader-centered discussions. It works well with subject-centered discussions. And it works particularly well with writer-centered narratives.

CONCLUSION

In this chapter, we have examined how to write good exits to articles. Everything that you have read can be transferred easily to writing conclusions for reports and memos. Remember, however, that the tendency is to overwrite the conclusion, to write beyond the stopping point. Only through practice and good editing will you develop a feel for knowing when to quit.

Following is an example of an exit from an article about computer technology. Notice that it is a forecast of future events:

> The Medibank Private Fund is embarking on a new $1 million+ media campaign that attempts to persuade Australians that "all health insurance funds are not the same." The campaign, which includes more than 130 television spots and insertions in Australia's three top magazines, will have ads tailored, for the first time, to young singles as well as to families.
>
> Capturing the healthier, and more profitable, young singles market will be crucial as Australia's health insurance and medical industries continue to change. The elimination a year ago of free hospitalization in Australia intensified competition among funds to be sure, but with medical costs having risen by 140 percent in the past year alone, Medibank Private and its competitors are finding the marketplace more challenging than ever.

Suggested Readings for chapters 9–12 appear at the end of chapter 12.

CHAPTER 12

How to Write the Lead

As I suggested at the beginning of Chapter 11, composing leads is the most difficult task in writing. Part of the problem is simply writer's block, that tendency for all of us to freeze up at the thought of putting words to paper. That problem will be dealt with in Chapter 24. But another major difficulty in writing leads is not knowing what an effective lead is, what it does, and how we can construct it. Writing leads last enables you to avoid many of these difficulties.

Effective leads bring the topic of a paper to the reader. In technical communication, they orient readers to the topic so that nothing that follows the lead is a surprise to the readers—at least not in terms of the subtopics covered. Now, perhaps we should consider exactly what I mean by this injunction. Although your topic itself might be revolutionary or surprising, you should not spring these new ideas in the main part of the article. In other words, you should not include information in the body that was not forecasted by the lead. That's what I mean by no surprises.

Just as several conclusions can be used for articles, a variety of possible leads can be written. We will consider these in the rest of this chapter.

THE OFFER OF SOMETHING NEW

This lead is used in many trade journal articles, as well as in the popular press, as an effective way to generate interest among readers. Almost all readers look to trade journals for the newest wrinkle in their specific field. If that is what you are presenting in an article, then you want to emphasize that fact in the lead. This lead works well with reader-centered discussions and with subject-centered discussions. It also ties in well with exits that forecast future events:

> America's competitive edge, dulled in recent decades by imports and outdated business tactics, is being honed anew. In both design studios and boardrooms, quality has taken on a new role—from that of lip service in advertising campaigns to new prominence as the ultimate business strategy. (Dana L. Gardner, *Design News*, Vol. 46, No. 3, February 12, 1990, p. 110.)

The article continues by describing the new strategy.

SUMMARY OF PAST DEVELOPMENTS

This lead (yes, it's yet another type of summary) serves to orient readers to a topic with which they might not be completely familiar. It is an attempt to show the main topic of an article as the natural outgrowth of what has occurred previously in the industry. Although it can be used with all three discussion organizations, it is a particularly effective way to start a subject-centered article:

> As the Voyager spacecraft head out of our solar system and set their sights on interstellar space, Bill Layman can take special pride. His design work on the space probes more than 12 years ago played a major role in a successful journey that yielded new scientific data and breathtaking photographs of the planets. (Carlton F. Vogt, Jr., *Design News*, Vol. 46, No. 3, February 12, 1990, p. 90.)

The article goes on to describe the design work.

COMPARISON OF THE OLD WITH THE NEW

This lead is good for a paper that is introducing a new development to an industry or scientific discipline. What better way to generate reader interest than to describe how your topic or product is better than what it is intended to replace. This way of subtly combining informative writing with advertising is something we all know is done, but it is something that is not talked about in public very much—at least in the public of technical writing. This type of lead also lends itself very well to use with subject-centered discussions, although it is usable with the other two organization styles as well. The following article uses this type of lead:

> "EUREKA! EUREKA!" Archimedes is said to have shrieked after he stepped into his bath and watched the water rise. In that instant, Archimedes realized that a body immersed in fluid is buoyed up by a force equal to the weight of the displaced fluid. Legend has it that he ran through the streets naked he was so excited about discovering this principle.
>
> Surprisingly, not much has changed in the intervening 2000 years. Water still rises when you get in the tub, and engineers still get some of their best ideas when they least expect it. The creative mind works in mysterious ways. (Tristram Korten, *Design News*, Vol. 46, No. 3, February 12, 1990, p. 138.)

The article continues by examining how creativity can be harnessed best in work environments.

CONCLUSION

In this short chapter, we have examined possible leads and what they can accomplish. This section is not a magic hat, however, from which writers may always pull out the right rabbit. The choice of a lead is closely tied to the writer's purpose, to the article's formality, and to its subject. Only careful analysis of these factors will provide the best basis on which to construct a lead.

Remember that the techniques discussed here apply mostly to articles. A formal technical paper almost always begins with a direct approach to the subject and the audience. The purpose statement introduction, which was discussed in Chapter 4, is an excellent way to introduce this kind of paper.

The following is a sample lead from an article about computer technology. Notice how the offer of something new is used to organize this lead effectively, without calling attention to the structure of the lead itself. This should be the writer's goal: an unobtrusive but always present structure in every communication.

> Even to people who do not program computers themselves, the value of computers as tools for many scientific, engineering, and educational applications has long been apparent. Until recently, though, the time spent learning how to use the computer, writing programs, or training a programmer often seemed to outweigh the benefits of computer use.

> Typically, someone with a series of calculations to do needs a set of specific, specially written programs to have a computer perform the calculations. Someone in this situation might adapt his needs to something "do-able" with commercially available programs. If that is impossible, he must either learn how to write the necessary programs himself or hire and explain his needs to a computer programmer.
>
> All these options have long lead times before a program can be used effectively. Also, unless the person has previous programming experience and writes it, he or she will probably have to compromise standards and use something that will not do exactly what is desired.
>
> Excel has solved these problems by providing popular mathematical programming capabilities that are built into their product.

Notice how this lead draws the readers into the topic, whets their appetites for more information, and leads them to the type of information they might expect (and should find) from the paper: namely, how the program can be used by people without programming backgrounds and how it is flexible enough to meet a variety of computing situations.

Chapters 9, 10, 11, and 12 examined the various components of writing articles about science and technology for media ranging from scholarly journals to newspapers. Scholarly journals usually have specific requirements for publication design and treatment, so reviewing those from the journal you are targeting is always a good place to begin. Writing about science for public media, such as magazines and newspapers, is much more open-ended. Following the suggestions in these chapters will result in articles that are well-organized and compelling—something that would please any editor. The example below (Figure 12.1) is an excellent application of what these chapters have presented.

ALLERGY PROMPTS REASSESSMENT OF
BEAUTY AND CLEANING PRODUCTS

By Erin Rehrig, Special to the *Telegram & Gazette*

Reading the list of ingredients in beauty products nowadays can be daunting. Most of us without a Ph.D. in organic chemistry struggle just to pronounce the ingredients, let alone understand what they do.

We often look for a product with the best price and 33% MORE FREE! on the label and throw it into our shopping carts. We blindly trust that the Food and Drug Administration ensures the ingredients in the products that we use every day have been thoroughly tested and are safe. But, what defines "safe"?

For about four months this past year, I suffered from a recurring full-body eczema that would have put the team on the TV show "House" to shame. To alleviate the symptoms, I tried moisturizers, lotions, creams and eliminating suspect foods from my diet, yet nothing worked. In fact, most of what I was doing was making the condition worse.

Finally, I was patch-tested for about 35 different chemicals by an allergy specialist and learned that the culprit was a widely used preservative called Methylisothiazolinone. I was relieved. All I had to do was simply buy new products that did not contain MI. Problem solved, right? Not so fast.

MI and it's derivatives (Methyl, Chloro- and Benzo- isothiazolinone) are used as preservatives to prevent microbial growth and are found in industrial biocides. They are also in almost every skin, bath, hair, beauty and cleaning product on the market, including so-called "natural products." (Challenge: walk down the shampoo aisle at the drugstore and try to find a product without them).

After my allergy results, I did an inventory; MI was an ingredient in my lotion, shampoo, conditioner, hair gel, body wash, hand soap, baby wipes, laundry detergent, all-purpose cleaner, furniture polish, carpet cleaner and dish soap. The FDA has determined that the levels of MI must be below a certain concentration to be safe. However, when applying 10 different products that all contain MI, there is bound to be an accumulation problem. Exposure to the allergen can be further aggravated when using leave-on products such as body lotion, hand cream and sunscreen.

Considering the allergist's office had a patch-test for MI, I was not surprised to find out that I am not the only person with an adverse reaction to it. In fact, Methylisothiazolinone has a history of being a potent allergen and was even named 2013's "Allergen of the Year" by the American Dermatitis Society.

Since the mid-2000s when inorganic preservatives were being phased out, MI has been used more often and in more products. Therefore, people are becoming increasingly sensitive to it. There are a growing number of cases of rashes and atopic dermatitis caused by MI exposure in the United States, Europe, Australia and Japan.

In fact, both the Japanese and Canadian governments have taken measures to curtail the use of MI in bath products. Recently in Australia, there has been some concern over the soaring number of infants with severe rashes caused by MI in baby wipes. There are also reports of air-borne MI released from latex house paint, leading to skin and respiratory irritations in Japan and the U.S.

Figure 12.1. Sample Scientific Article. (Rehrig, Erin. "Allergy prompts reassessment of beauty and cleaning products." *Worcester Telegram & Gazette*, April 2, 2014. Used by permission of Erin Rehrig.)

SUGGESTED READINGS FOR CHAPTERS 9, 10, 11, AND 12

Gastel, Barbara, and Robert A. Day. *How to Write and Publish a Scientific Paper,* 8th ed. Santa Barbara, CA: Greenwood, 2016.

Schimel, Joshua. *Writing Science: How to Write Papers That Get Cited and Proposals That Get Funded.* New York: Oxford University Press, 2011.

Silvia, Paul. *Write It Up: Practical Strategies for Writing and Publishing Journal Articles.* Washington, DC: American Psychological Association, 2014.

PART IV

How to Write Specific Documents

CHAPTER 13

How to Write Specifications

For everyone involved in the design phase of high-technology industries—engineers, technicians, programmers, architects—specifications are among the most important documents to be read or written. They dictate design.

Obviously, the audience for specs is highly technical, but this does not mean that the audience necessarily shares the knowledge of the writer. This lack of common knowledge is particularly true when specifications cross major professional lines within a high-tech industry; for example, a specification written by marketing people for engineers, or even a spec written by one type of engineer for another type of engineer, may not be useful if the writer assumes too much of the reader. So, even though the language of specifications is precise and even though the treatment of the subject is thorough, great care must be taken to ensure that the specs can be used by the intended audience. Nothing wastes more time within an organization than having to figure out poorly written specs. The situation is even worse when work has to be undone or redone because of bad specs.

TYPES OF SPECIFICATIONS

For our purposes here, specs can be subdivided into four types (other subtypes exist, but that is a more specific treatment than we need in this book; for additional information, look at the list of suggested readings at the end of this chapter):

- requirements specifications
- functional specifications

- design specifications
- test specifications

Although the names for these specifications may differ from company to company, each type can be found in technological companies that design, produce, and market products. All four types will be discussed in this chapter.

Requirements Specifications

The requirements specification should be the first step in the design phase for a new product or for a product that is being updated or changed. Often, people in the marketing division of a company are responsible for determining what the market desires. At the lowest common denominator, this is why market research is done.

The result of market research is the requirements specification. In it, marketing professionals attempt to specify what the market is looking for, what technology customers would find useful and would like to have. A former student of mine who now works for Apple, and writes these types of specifications, refers to this process as "writing corporate fictions." Requirements specifications specify something that does not exist, and because they are often written by people who are not design engineers, they are likely to be the most general of specification types. They can do no more than provide marketing's best effort to describe a profitable product.

Even so, they are extremely valuable. They provide the design group with a place to start. Often, they will contain enough information so that readers can see relationships to past technology. To do that, requirements specifications should contain the following as a minimum:

- Product Definition—As accurate a description as can be written by marketing about the desired product. It should answer the question "What is it?"
- Functions List—A description of what the desired product should be capable of doing. It leads to the next type of specification.
- Cost—A ballpark estimate of what the desired product should cost to be competitive in the marketplace.

Functional Specifications

In many organizations, after the requirements specifications have been written, a group is formed to study the desired project. For industries in which a product is the combination of the physical product and the software that runs it, this study is often subdivided into hardware functions and software functions, and it leads to the writing of both hardware and software functional specifications. These specs form the basis for the highly precise design specifications.

Hardware functional specifications as a rule contain the following, often in this order:

- Functional Description of the Product—A precise description of the purpose, use, and operation of the product. It might reference related documents; discuss user, performance, and compatibility requirements; and present enhancements or options.
- Configuration Specification—Specifies such things as how the product's components are to be interfaced with each other and with other available products.
- Electrical Description—The specification that describes the electronics that will be used to accomplish the product's capabilities.
- Physical Characteristics—This section contains a precise description of each of the product's components.
- Standards—Specifies how the desired product will fit into existing company standards.
- Environmental Requirements—A description of how the product will be used and under what conditions.
- Diagnostic Requirements—A description of the testing and evaluation requirements for the product.
- Power Requirements—Describes what sort of power source will be required to operate the product.
- Cost Target—Establishes what the product should cost the consumer.
- Maintenance Cost Target—Establishes what the expected maintenance costs are likely to be.
- Resource Requirements—Specifies what resources will be needed to design the product.
- Documentation—Outlines the necessary documentation for the product. In some organizations, this is referred to as "the information product."
- Risks—A discussion of the risks inherent in pursuing the design, development, and production of the product.
- Assumptions—Describes the underlying assumptions that can be made about the product and the processes of designing, developing, producing, and marketing it.
- Unresolved Issues—Even in documents this thorough, issues can remain unresolved. These issues should be presented and discussed.
- Glossary—Ensures that readers from a variety of technical and nontechnical backgrounds can understand and use the specification.

Software functional specifications are similar, but enough differences exist to warrant examining them. Sections that repeat sections of the hardware functional specification will not be explained again. Software functional specs usually contain the following:

- Functional Description of the Product
- Product Features—This section describes the capabilities of the product in detail.
- Environment
- Dependencies—This section is an elaboration of what the implementation and use of the software will depend upon.
- Physical Characteristics
- Risks
- Assumptions
- Cost Target
- Maintenance Cost Target
- Resource Requirements
- Documentation
- Glossary

The level of detail increases dramatically between the requirements specification and the functional specifications. That detail will become even more specific in the design specifications.

Design Specifications

Design specifications are based on functional specifications. The goal of design specs is to provide a detailed design of each of a product's features. These specifications are best begun before the design process starts and updated while the design process continues. Design specifications are later used as a basis for test plans and user documentation. As with functional specs, design specifications can be divided into hardware design specs and software design specs.

Hardware design specifications generally contain some version of the following components, often in this order:

- Introduction—A three-part purpose statement introduction that explains the need for the product, lists the specific features of the product, and forecasts the contents, organization, and use of the design specification.
- Applicable Documents—A list of the documents that contain information pertinent to the product. Such a list is absolutely essential to the

technical writers who will be producing the documentation manuals for the product.

- Functional Description—A detailed description of the product's functions, what it is designed to do, and how it is designed to do it. Various graphics are used extensively in this section, but make sure that they are not a substitute for clearly written text.
- External Interfaces—Specifies all interfaces that apply to the product.
- Detailed Design—Details the design of individual aspects of the product's functions. It is the most detailed section of the design specification.
- Programming Considerations—Describes all aspects of the hardware with which a programmer would come into contact.
- Power Requirements—Describes the assumed power requirements for the product.
- Reliability—Explains how reliable the product is designed to be and what is to be expected of it with regard to service and maintenance.
- Diagnostic Considerations—Describes the testing and evaluation requirements for the product.
- Standards—Explains how the product will fit into existing company standards.
- Environmental Requirements—Describes the operating conditions that are assumed for the product.
- Glossary—Defines potentially unfamiliar terms for a wide range of readers who might have to use the specifications.

The software design specification is also similar to the hardware design specification. But as was the case with functional specifications, here, too, are differences that need to be explained. The software design specification is also used as a basis for testing and as the source for user documentation. To meet these two purposes, software design specifications should contain the following:

- Introduction
- Applicable Documents
- Functional Description—This section should be subdivided into however many functional features the software has.
- General Design—The design section details the way in which the software design objectives are met. It is the most detailed section of the software design specification, including such material as data structures, data flow, program relationships, and so on.
- Memory Requirements, Performance, and Restrictions—Details how the software fits into and uses computer memory. It assesses the performance of the software and presents any restrictions that might apply.

- Product Requirements—Discusses such matters as security, usability, and installation and maintenance requirements.
- Test Strategy—The test section presents any helpful suggestions that could be used in developing a test plan for the software.
- Deviations from Functional Specifications—Sometimes, changes are necessary. This section describes them.
- Interface
- Glossary

User documentation for high-technology products is universally lambasted for being unreadable and unusable. Much of the problem lies in specifications, which later become source material for technical writers. If high-tech professionals would take the time to write and appreciate the importance of thorough design specifications, to anticipate one of the eventual audiences of these specs (technical writers and the users), and to follow a version of what has been presented thus far in this chapter, user documentation would improve drastically.

Test Specifications

Before a product can be marketed, it must be tested to see how well it will perform under market conditions. This procedure also should be specified consistently across a company. To do that, test specifications should contain at least the following:

- Introduction—A three-part purpose statement introduction that helps to describe the purpose of the tests and to forecast the contents of the specification.
- Applicable Documents—These documents might describe test procedures on similar products that have been designed and developed in the past.
- Description of the Unit to Be Tested—Identifies and describes, thoroughly, the test unit.
- Testing Method—Provides a step-by-step description of the testing procedure.
- Precautions—Details any special care that must be taken in the testing phase. The same degrees of precaution apply here as they do for procedures. See Chapter 15.
- Glossary—Here the glossary defines potentially unfamiliar terms for the people who will be conducting the tests.

CONCLUSION

In this chapter, we have examined four broad categories of specifications common to high-tech industries, especially those which design and produce products combining hardware and software (which, of course, in the 21st century is virtually ubiquitous). Minor differences will occur from company to company and from industry to industry, but one universal is a high degree of accurate, exhaustive description. This accuracy is both necessary and desirable. Too often specifications are hastily written with no thought for their eventual use. Writers see only the present and figure that the limited group of in-house readers can work through turgid specs. This assumption is untrue, and it forces unnecessary work upon all readers, leading to projects that fall behind schedule and run over budget, as well as to user documents that no one can use. At a time when more and more companies are pointing out that their information products can affect the marketing success of a company's products, clear specifications are vital.

SUGGESTED READINGS

Schauer, Jane, ed. *How to Write Specifications.* Kraev Publishing, 2009.

Timms, Linda. *SAP: How to Write an Interface Functional Specification.* Bloomington, IN: AuthorHouse, 2012.

Timms, Linda. *SAP: How to Write a Report Functional Specification.* Bloomington, IN: AuthorHouse, 2012.

CHAPTER 14

How to Write Procedures

About 40 years ago, while I was in graduate school and suffering from the delusion that I was good with my hands, I bought my wife a bureau in a box. Like most people when they start work on a kit, the first thing I did was open one end of the box and dump all the contents out onto the floor of our apartment. The last thing that fluttered out was the instructions. The first thing they said was, "Remove contents from box in an orderly fashion." The wisdom of this advice was immediately apparent. There was no parts list, no handy little plastic bags containing the hardware, just a lot of wood and metal scattered on my floor. The next thing I noticed was that the instructions had only three steps. At least 40 objects were sitting on my floor, and even though the project was supposed to be simple (one that could be done in 3 hours, or so the box had said), I had my doubts. So, I did what most people do when they find themselves in this predicament: I threw the instructions away. Three days later, something resembling a bureau had grown in our apartment, but one small problem remained. The last piece of wood, which looked as if it ought to be a restraining piece fitted across the back of the bureau, was four inches longer than the top of the bureau. Nothing suggested it should fit diagonally, a solution that would have looked odd anyway, and I did not want the extra four inches sticking out into the room to grab people as they walked by. So again, I allied myself with kit builders everywhere in this kind of predicament: I got my saw, sawed the excess off, and nailed the piece in place. The bureau stood up and enhanced our home for years.

For several years, that was the worst set of instructions I had ever encountered. But in the late 1980s, I found the all-time winner. I had contracted for an addition to be built onto my house. The contractor said I could save $4,000 by adding a woodstove instead of the fireplace my wife and I wanted.

This argument was persuasive, so I went out and bought a woodstove I could install myself. I knew there was going to be some problem when the woodstove dealer brought the product to my Jeep using a forklift. I told him that was nice, as he placed the woodstove in my car, but that I did not happen to have a forklift at home. No problem, he responded; simply take the stove apart in the box, carry each piece into the house, and reassemble it there. The logic of this advice was unassailable, so I left for home with a brand-new, shiny, efficient Norwegian stove that cost me just under $1,000. I did what the dealer said, and after I had scattered the woodstove around my new room, I located the instruction manual at the bottom of the box. The first five pages were in Norwegian. The next four were in Swedish, the next four in German, and the next three in French. The last page contained the English translation, which consisted of just two sentences: Step 1. Install woodstove, and Step 2. Have fire inspector check installation. Chalk one up for technical-writing brevity.

These stories do serve a purpose in this chapter. They identify the single most common problem of poor procedure writing: procedure writers who have no idea whom they are writing for and under what conditions the unfortunate user will be doing the procedure. Consequently, the worst technical writing tends to be documents in the how-to category. I have even heard people say that they are convinced the persons who write procedures have no idea how to perform them. While I would not go that far, such an attitude does suggest a public relations problem for those of us who write procedures. Our informational product often does not inspire confidence that even we know what we are doing. That is reason enough for considering how to write procedures. This chapter, then, will do just that: examine a method for planning, writing, and testing procedures.

PLANNING FOR PROCEDURES

Before readers can be told how to do something, they must be told what the something is and what its purpose is. In other words, give readers a rationale for doing the procedure. Doing so suggests to readers that you have considered them and what they are up against in their day-to-day activities, and that you don't want readers wasting time trying to figure out what you want them to do. This consideration has an obvious positive effect: readers are more likely to follow the procedure accurately.

You can further enhance the reader's sense that you considered them by considering their different needs for information. Nowhere is the difference between personality-based preferences for information more important than in the writing of procedures. Sensing readers prefer a detailed, step-by-step chronology while intuitive readers need only overviews and a general sense of what they are going to do. Be sure to provide readers with both types of information.

Procedures, regardless of the format, may do any or all of the following:

- give steps for operating something
- give steps for assembling something
- give steps for troubleshooting, repairing, adjusting, or maintaining something
- give steps for unpacking or shipping
- give steps for ordering parts, optional attachments, and so on
- furnish a parts list
- teach skills, as in a training manual

The important fact that you must realize when planning and writing procedures is that you become a teacher. You are the expert, and it is your responsibility to impart knowledge to readers who know less about the subject than you do. If the following statement is true of any type of writing in the computer industry, it is especially true of writing procedures: The written product has value only insofar as it can be used.

One of the ways you can ensure the usability of your procedures, and a way you can instill confidence in your readers that you know what you are talking about, is through the language you use. Use the command voice, rather than the descriptive voice. An example of that is the preceding sentence ("Use the command voice . . ."); sentences that begin with verbs are in the command voice. Notice that the descriptive voice (for example: "You should use the command voice") suggests that there might be alternatives to what you want done, that the reader only "should" use it, not that he or she absolutely has to use it, no questions asked. The command voice is the voice of authority. As long as you don't fail your readers with inaccuracies, the command voice is convincing.

Vocabulary is also important in procedures. The procedures must be written in the terminology of your readers. Impressing them with your erudition or vocabulary is worthless. If readers are to follow a procedure accurately, they at least have to be able to understand the individual steps. This necessity is just one more indication of the importance of audience analysis and adaptation. As a writer of procedures, you must know who your readers are and what level of expertise they have attained with regard to the subject. Depending on what you know about your readers, you may have to define terms and explain unfamiliar concepts. Clarity is your goal, and that means explicitness. If, after you have considered your audience, you still are not sure what to include, aim on the side of too much information. Too much information may inconvenience some readers; it may aggravate others, but it's a small price to pay. Too little information means the procedure can't be done.

Planning for procedures begins before you start to write. First, you have to know your subject. Find out how the subject came into being. If it is something that has been designed (new software, for example), find out why. What needs will it fill? Who will it be marketed to? Examine the product's specifications. You should also examine the subject yourself. Play with it; use it. If it is something that must be assembled, assemble it. If it is a programming procedure that must be followed, follow it. See if it works. Find out its purposes, who will use it, and under what conditions. What are its good and bad features? Even though your company may not want you to say, "Here is where our product is likely to break or malfunction," you still need to know these things. You can deal with these problems tactfully in a troubleshooting section.

If your subject is complex, divide it into parts. These parts will later become sections in short procedures or chapters in manuals. You may even have to subdivide the subdivisions. These divisions, marked by headings, enable readers to see the important parts of the procedure and how they interrelate. Naturally, how you divide up material depends greatly on the design you are using for the information. For example, modular design, which Edmond Weiss describes in detail in *How to Write Usable User Documentation,* requires that you subdivide information so that it fits into very specific sections of a document, usually no more than two facing pages and often only one page. Chapters and sections are less restrictive in terms of the physical requirements of document layout, but information in them must be highly organized, too.

If your subject lends itself to things that might go wrong when the procedures are not followed accurately, make a list of precautions your readers should know. The following are headings that have widely accepted meanings.

DANGER—reserved for steps in a procedure that could lead to serious injury or loss of life if readers do not know what they are doing.

WARNING—used for steps that could result in damage to the product if the procedures are not followed accurately.

CAUTION—applied to steps where faulty results could occur if the reader does not follow the procedure correctly.

NOTE—used to alert readers to potential problems and for making suggestions that would make the reader's work easier.

Once you have analyzed the subject, considered the projected audience, made an outline, and listed the precautions that will apply, you are ready to write the procedures.

WRITING PROCEDURES

Writing procedures begins with determining the purpose of the procedures for the intended audience. This process includes choosing an explicit title that includes the date and all appropriate models and operations to which the procedures apply. It also often includes writing a short purpose statement introduction. This step is the most often overlooked aspect of procedure writing. Many writers assume that since they are telling readers what to do, nothing is served by telling them why to do it. This belief is a dangerous fallacy because it assumes readers will automatically see the wisdom of doing it your way. Not so! Once these concerns have been met, you will have to choose a format that suits your purpose and that communicates the information efficiently to your readers. A variety of choices are available. Figure 14.1 shows the VOS (Visual Organization Systems) for procedures.

TESTING PROCEDURES

The final task procedure writers have to concern themselves with could be called quality control. Usually, we think of quality control as something that

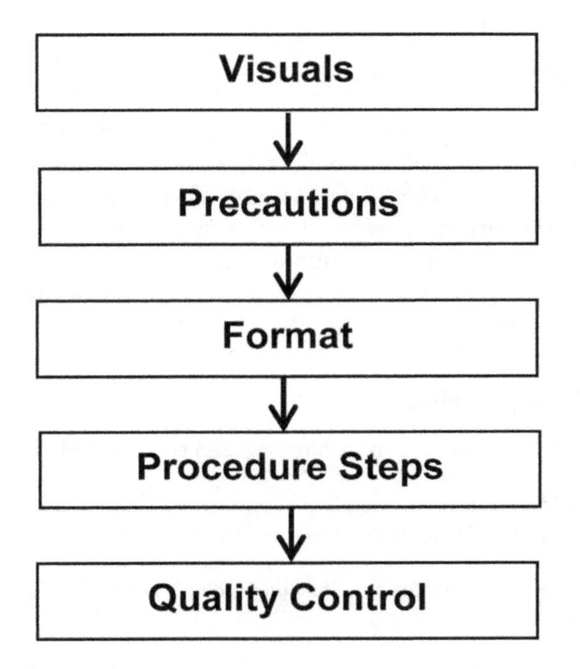

Figure 14.1. VOS for Procedures.

applies to production items, something that industrial engineers do. But quality control also applies to written documents, or information products, and it particularly applies to procedures.

To check the quality of your procedures, make sure that two types of people read and evaluate them before the procedures are made public. Have an expert in the technical area of the subject review the procedures for accuracy; this reading is often referred to as a technical edit. But don't stop there. Have a person who represents the ability level of the users read it to see if the procedure can be understood. This step is known as a usability edit, and it is vital to the effectiveness of your final product.

After these edits, you might want to beta test the procedure in the environment in which it is intended to be used with persons intended to use it. This practice is commonly used for evaluating the functions of newly developed hardware and software. Since the first edition of this book was published, it has become increasingly common to extend beta test environments so that they include documentation. Some companies, such as IBM, have gone so far as to videotape users dealing with the hardware or software product and the documentation, making comments about their trials and tribulations as they go along. Aside from the fact that these videotapes are hilarious, they are extremely useful in improving both technological and informational products.

CONCLUSION

What I have done in this chapter is give you a procedure for writing procedures. As is the case with all procedures, the success of this one will be determined on the basis of how well you are able to write procedures for a variety of audiences. Remember, however, that this chapter is a procedure for developing a skill. Just as a potter, through practice, becomes competent at throwing pots, you, through just as much practice, will improve your procedures.

SUGGESTED READINGS

Damelio, Robert. *The Basics of Process Mapping,* 2nd ed. New York: Productivity Press, 2011.

Muchemu, David. *How to Write Standard Operating Procedures.* CreateSpace, 2010.

Page, Stephen. *Establishing a System of Policies and Procedures,* 7th ed. New York: Page, 2012.

Peabody, Larry. *How to Write Policies, Procedures, and Task Outlines,* 3rd ed. Lacey, WA: Writing Services, 2006.

CHAPTER 15

How to Write Proposals

Strictly speaking, a proposal is a sales piece, a communication designed to obtain work, funding, a go-ahead on a project, and so on. Its ultimate goal is to identify a need on the part of the audience and to outline ways that the writer, or the group of people the writer represents, can satisfy that need. For proposals to be successful, writers must convince the audience that they can do something for the audience, that it needs to be done, and that the writers or proposing group can expect some sort of recompense for having done it.

As it appears, this type of writing is not easy. First, proposals must invite readership if they are to be successful. Readers must want to go beyond the first sentence of the introduction. That is one reason why three-part purpose statement introductions are particularly important to proposals. Such introductions get the reader oriented quickly to the aims of the document.

Besides the introduction, however, other steps can be taken to invite readership of proposals. Most importantly, make sure that the proposal is easy to read and understand. Accordingly, you should try to do some, if not all, of the following:

- Consider the needs and levels of understanding of the audience (i.e., analyze your audience).
- Use a simple format (more about that later).
- Make sure that the final draft is clear and legible (high-quality print, attractive and accessible layout).
- Keep paragraphs and sentences reasonably short (shorter in proposals than in analysis reports, for example).
- Use headings (even in informal proposals [see Chapter 20]).
- Use the active voice whenever possible (more on that in a later chapter).

Finally, realize that readers read proposals in order to reject them. That's right, and it may come as a shock to anyone who thinks that readers are eagerly awaiting their analysis and opinions. This approach to reading proposals is especially true of proposal evaluators in large, competitively bid projects. You will want to provide skimming cues for intuitive readers, as well as sufficient details about the project for sensing readers. Make it easy for people to judge your proposal by supporting it with logical reasons (for thinking evaluators) and with how it fits into their value systems (corporate, national, etc.) for feeling readers.

In this chapter, we will look at ways of writing successful proposals. Specifically, the following will be examined:

- proposal formats
- title pages for formal proposals
- proposal introductions
- discussion sections
- conclusion sections
- Gantt chart time schedules

Figure 15.1 depicts a method for organizing your proposals.

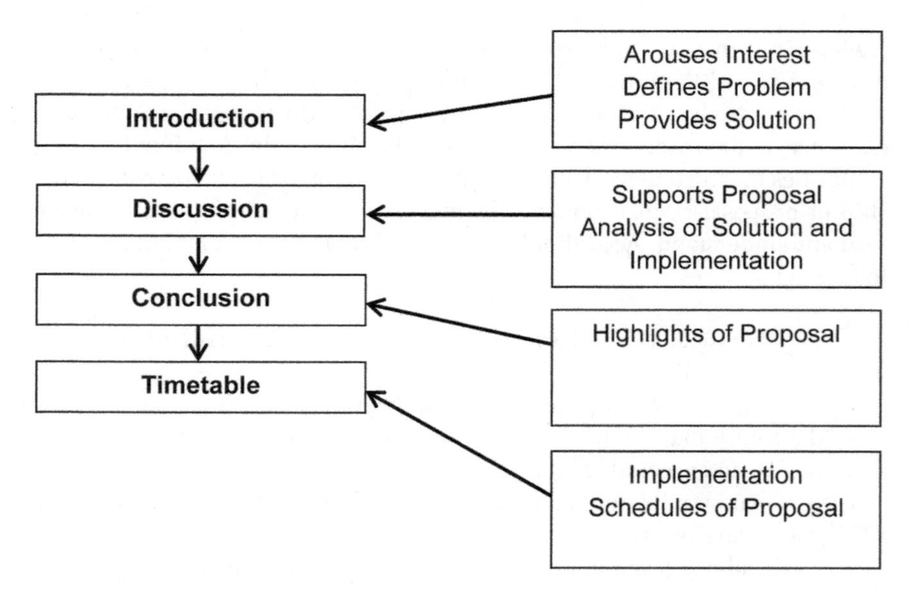

Figure 15.1. VOS for Proposals.

PROPOSAL FORMATS

The following proposal formats are general guidelines, not prescriptions. More than likely, the company you work for (if it is a large company) has style guidelines for reports of this type. The material presented here, however, is easily fitted to your company's format. If your company does not have a style guide, the information presented in this chapter, and in the rest of this book, can be used to help design a company style guide. In embarking on such a project, remember that the two most important aspects of report format design are the needs of the audience and the purpose of the writer and the company.

All the formats presented in this section are for informal proposals—those without title pages, letters of transmittal, tables of contents, prefaces, acknowledgments, and the other trappings of formal reports. The following format is designed for problem-solving proposals.

Format for Problem-Solving Proposals

- Introduction—A three-part introduction that defines the problem, briefly describes the proposed solution, and briefly outlines the benefits of adopting the solution.
- Discussion—An exhaustive analysis of the solution, including its benefits and challenges.
- Resources—What is needed to implement the solution, including what is available and what must be obtained.
- Costs—Dollars and cents (how much will the solution cost in dollars and cents?).
- Personnel—The people who will implement the solution and the people affected by it.
- Schedule—A Gantt chart for implementing the solution (more on this a little later).

The following format could be called an executive summary proposal, since it aims to satisfy the needs of the reader who does not have the time or the inclination to read the entire project. The important criterion for a successful executive summary proposal is to include enough of the right kind of information in the summary for the reader to be convinced that you can do what you propose to do.

Format for Executive Summary Proposals

- Summary—A brief but complete discussion of what is at issue and what is being proposed. The idea is to give the executive reader enough information to make a decision at this point.

- Introduction—A transitional introduction that focuses on the background to what is being proposed and why. Orient readers here, preparing them for the detailed discussion that follows.
- Discussion—A detailed analysis of the proposal, its benefits, challenges, and implementation.

Note that this format is essentially two statements of the proposal—a basics version for the executive audience and a detailed version for the technical audience.

The final format to be considered here is an extension of the first two. Whereas either of the previous proposal formats could be used for memo proposals or short informal proposals, the following format is generally used for longer proposals, proposals that might be used to suggest a project of some sort. Formal proposals usually are based on some version of this format, as well.

Format for Project Proposals

- Introduction—A three-part purpose statement that orients the audience to the background which led to the development of the proposal, the specific goals of the proposal, and the purpose of the report in terms of its readers.
- Statement of the Proposal—A complete description of what is being proposed.
- Management Section—A comprehensive analysis of the budgetary and personnel responsibilities included in the proposal.
- Technical Section—A comprehensive analysis of the research, engineering, design, development, and material aspect of what is proposed.
- Costs—A detailed projection (often visual) of proposed costs.
- Schedule—A Gantt chart.

Note that in this format, writers often supply their own subject subheadings for shorter subsections within the major divisions of the proposal.

Some companies may require that proposals for large research and development projects be written as three separate but related documents:

- technical proposal
- management proposal
- cost proposal

The three proposals are merely an extension of the formal project format presented earlier. Appropriate front matter (title page, summary, introduction) and back matter (schedule, appendixes, resumes) would be added to the technical proposal, management proposal, or cost proposal depending upon which type of proposal was being written.

Realize, too, that these three elements of the proposal, whether they are written as one document or as three, make up the parts of the proposal that are most closely evaluated. In many cases, the proposal that offers the most or best for the least cost will be the winning proposal. But in many more cases, deciding what is best is not a clear-cut decision for those who evaluate proposals. Because of this lack of clear choices, the writer has considerable influence by making sure that the argument in the proposal is clear and persuasive. As I mentioned in the beginning of this chapter, proposal evaluators read proposals not to see where they are right but to see where they are wrong.

The next few sections will examine how to make proposals read right.

WRITING PROPOSALS

Title Pages

Title pages are often included on informal proposals and are always included on formal proposals. A title page should contain enough information so readers can easily identify the contents of the proposal, its authors, issuing organization, and date of issue. It should be visually balanced, imparting a sense of order to the reader. The not-so-subtle suggestion here is that the orderly appearance of a document reflects the orderly mind behind it.

Introductions

Introductions to proposals must orient readers to the contents and the purpose of the proposal, just as an introduction would do for any other report. The important difference, however, is the purpose of the writer. In most reports, writers are only relaying information. In proposals, they are convincing the audience that what is being suggested should be accepted, adopted, bought, or whatever. Persuasion becomes an important aspect of the introduction to a proposal.

Discussion Sections

In the discussion section of a proposal, writers describe the proposal, its benefits, and its implementation in full. Persuasion is no less important here.

One way to enhance the readers' understanding of the discussion is to make sure that the organization of information follows some clearly identifiable pattern.

Depending on the format and the intentions of the proposal writer, the discussion section of a proposal may do any of the following:

- In problem-solving proposals, it may present an in-depth description of the solution. This section would include subsections on the benefits of the proposed solution, on any anticipated difficulties in implementing the solution, on the amount and type of work required to implement the solution, and so forth.
- It may present a detailed description of what was highlighted in the summary of executive summary proposals.
- In long project proposals, this discussion is usually subdivided into subsections centered around the managerial or technical responsibilities for what is proposed. Because more information has to be communicated to the reader, the subdivisions are necessary.

Following is an excerpt from a problem-solving proposal discussion. Notice the clear organization of what is proposed, followed by supporting statements.

HARDWARE PURCHASES

Unfortunately, we do not have a closet full of surplus computer hardware. However, the computers we do have will provide an excellent basis for further expansion.

As a result, I recommend the following:

1. That we no longer use desktops and that we adopt laptops instead.
2. That all computers be equipped with Bluetooth capability.
3. That we purchase a Wi-Fi network system that will allow us to keep all our computers under one control and linked throughout the system.
4. That all future hardware purchases be compatible with our present computers.
5. That we offer in-service programs and summer workshops for teachers.
6. That we appoint a director of computer education.

I also recommend that we purchase the following:

1. One Lenovo ThinkServer TS140.
2. 16 Lenovo ThinkPadX computers.
3. One Cisco Meraki MR18 wireless access point.

The reasons for my recommendations are as follows:

1. If we upgrade to laptops, then we will be able to buy better educational software.
2. A Wi-Fi system will allow a teacher to present the same skills to students throughout the school.
3. We have the technical expertise in the school to install everything.
4. Lenovo can be contracted at minimal cost for teacher workshops.
5. In order to develop curriculum, we must have in-service workshops to put together a course of study and to order new software.
6. A director is needed to oversee the entire program. A director can order new software and be responsible for all the equipment.

Almost all the producers of quality educational software have products suitable for laptop use. The Wi-Fi network system, properly used, means that a teacher can better use his or her time to focus on educational experiences. This ability will save considerable time.

Because Lenovo offers excellent services, they will be an excellent resource for ongoing maximization of the computer purchase. This is the only company that offers these range of extended services.

Finally, a full-time director will be able to put together loose details and direct future development. This direction is important because the classroom teacher does not have the time to participate in the running of a computer center and to maintain the responsibility for a classroom.

Conclusions

Conclusions to proposals serve two purposes: to put the reader in a positive, receptive frame of mind, and to suggest to the reader how to obtain the benefits of the proposal. They should be brief and to the point. Benefits should be highlighted, and required actions should be stated explicitly. Good conclusions give readers a way to assess the proposal, to see if the author is on the right track. They also force authors to think beyond the proposing step, to examine the expected results of a successful proposal. By forcing the writer to think ahead, conclusions make time scheduling more straightforward.

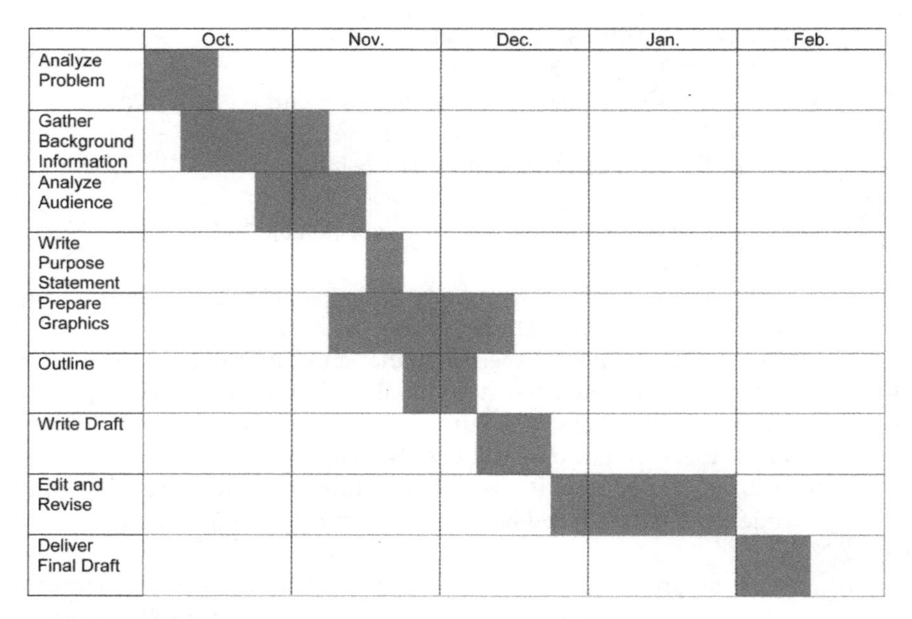

Figure 15.2. Gantt Chart.

Gantt Chart Time Schedule

All proposals should show the audience how long it will take to enact a successful project once it is adopted. One of the best ways of depicting a schedule is to use a graphic known as a Gantt chart. A Gantt chart takes a proposal and divides it into tasks to be accomplished and the dates by which they will be accomplished. It is a version of a common bar graph (explained in Chapter 7). Figure 15.2 is a sample Gantt chart of a documentation task. These charts can be used, however, for any process that allows tasks to be subdivided over a period of time.

CONCLUSION

In this chapter, we have examined some of the aspects of writing successful proposals. Realize that no one formula can guarantee successful proposals. Writing them takes a willingness to shape information so that it meets the readers' needs, so that it fits into the requirements of a company's style guide, so that it satisfies the author's purposes, and so that it is convincing to readers. Combining all these elements is a tall order, but it is not impossible. Just remember to make it as easy as possible for readers to evaluate and accept your proposals.

SUGGESTED READINGS

Forsyth, Patrick. *How to Write Reports and Proposals.* London: Kogan Page, 2016.
Forsyth, Patrick. *Persuasive Writing for Business.* Bookshaker, 2014.
Mugah, Joseph. *Essentials of Scientific Writing.* Bloomington, IN: AuthorHouse, 2016.
New, Cheryl Carter, and James Aaron Quick. *How to Write a Grant Proposal,* 3rd ed. Hoboken, NJ: Wiley, 2003.

CHAPTER 16

How to Write Analysis Reports

Analysis reports are called by different names from company to company—formal reports, project reports, final reports, or job-end reports. I have chosen to call them analysis reports because that is the one characteristic they have in common: they all analyze a completed project to assess its success or failure. As such, these reports are an important aspect of documentation within the scientific and technological companies industries. Filed away for future reference, they can be referred to by people working on the design of later products or by people writing user documentation.

In this chapter, we will look at a general approach to designing analysis reports. The important thing to remember, however, is that no report format is perfect. Company documentation standards attempt to resolve the issue by prescribing a format into which all analysis reports are poured. But even this prescribed format does not work all the time. Report design should be flexible enough to meet a variety of writer purposes and audience needs. For example, you could consider fitting your analysis of a project into the overall value system of the corporation. Doing so could conform it to certain type preferences for maintaining order and past corporate values. Or, you could provide for a two-tiered readership by including summaries, headings, and overviews for skimming for intuitive readers and logically organized details for sensing readers. Since these reports are the basis of analysis, make sure that yours is rigorously logical. You will want to prove your judgments, especially for the thinking types in your audience.

The format presented in this chapter can be used as a basis for your own report design or as a starting point for developing report standards for your company. In addition, it can be used as a general guideline for organizing a formal technical or scientific paper since the difference between analysis

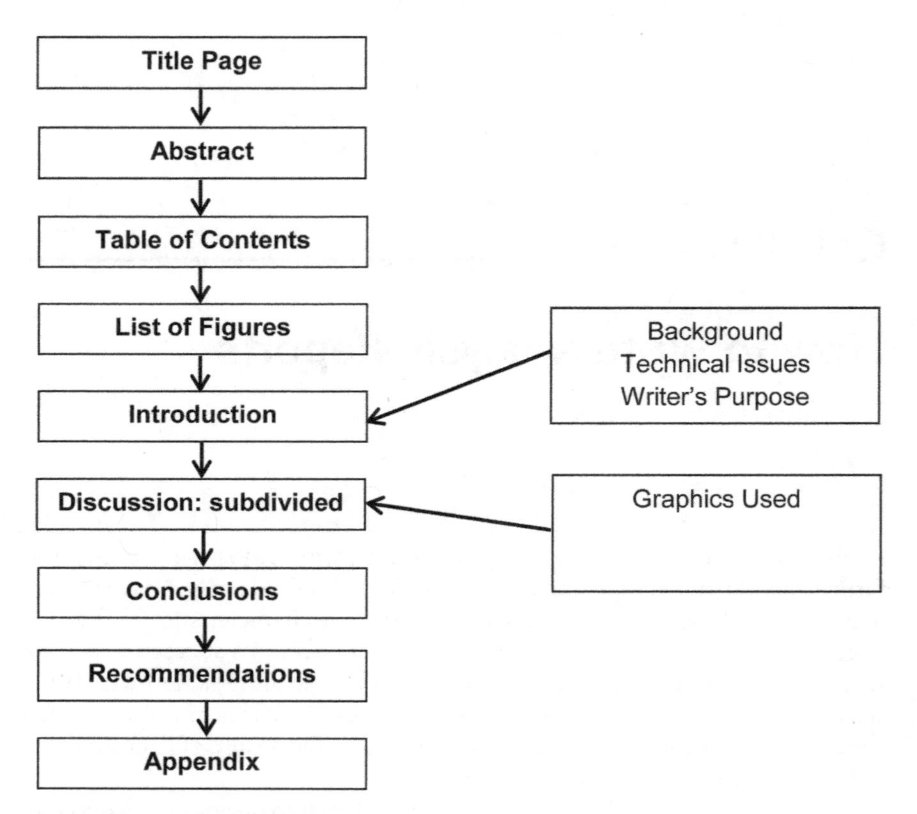

Figure 16.1. VOS for Analysis Report.

reports and technical papers is often simply the medium of publication. Figure 16.1 shows this format.

Each aspect of analysis reports will be discussed in detail, with the emphasis on options available to the writer.

SECTIONS OF A REPORT

Title Page

A title page should be designed with visual order in mind. It should be balanced from top to bottom and from left to right. It should provide enough information so that readers can tell what the context of the report is and what the report is about.

Titles should be kept relatively short, no more than 10–12 words. Avoid one- or two-word titles, however; they suggest that you are going to tell everything there is to know about the subject. And in most cases, that is not what you intend to do. Two-part titles can be a little longer. They are useful because the first part is general, providing the context; the second part is specific, indicating the main point of the report or the most important aspect of what was done. The following is an example of a two-part title:

"SERIOUS GAMES: Games whose purpose is not limited to entertainment"

Abstracts

Abstracts are condensations of entire reports, focusing on the main issues: what was done, what was found out, and its significance. Abstracts are self-sufficient. Their use in many companies involves taking the abstract from the analysis report, photocopying it, and circulating it through a distribution list or maintaining it on a website for readers who will then decide whether they need to order the complete report. This arrangement places a lot of responsibility on abstracts. They have to work for busy readers who do not have time to read the entire report and for researchers who want to know if the report is useful. To do these things well, abstracts must be informational.

Informational abstracts relate important aspects of the report. They should not contain words such as "described" or "presented." If they do, they are likely to be descriptive abstracts (little more than prose tables of contents) that tell what the report is about, not what information is important. The following is an example of an informative abstract:

> PLOTFILE is a program written for Concord Sciences Corporation to solve their graphing problem. Specifications were made by the president of the company and followed with two exceptions. The program, now in use, also has additional features. The user provides information through interactive commands to specify desired plotting formats. With flexible logarithmic and linear capabilities, the program then draws the desired axes or grids and plots the set(s) of points read from the data file.

Notice that the important features of the program are highlighted in this abstract. Readers need to know that this information is in the report. Compare this informative abstract with the following example of a descriptive abstract.

> This user's guide for the Motorola MG7550 Wi-Fi Cable Modem describes commands, statements, functions, and the modem. Online setup is provided, with no call to Comcast needed.

You can tell what the manual is about, but you can't tell if you can actually use the information.

Table of Contents

The table of contents provides an outline of analysis reports for readers who do not wish to read the entire report or flip through it looking for the section that contains what they want. It should be made up of the headings and subheadings of the report, word-for-word with the accompanying page numbers. It should also be orderly and easy to read. The following is an example of a table of contents from an analysis report:

1. ABSTRACT...........1
2. INTRODUCTION...........2
3. TECHNICAL ANALYSIS...........3
   3.1. Modem & router gateway standards...........3
   3.2. Industry standard AC1900...........5
4. SYSTEM IMPLEMENTATION...........6
   4.1 download channels...........6
   4.2 Wi-Fi Pairing...........7
5. RESULTS...........10
6. CONCLUSIONS...........11
7. RECOMMENDATIONS...........12

APPENDIX A—DESCRIBING FUNCTIONS...........13
APPENDIX B—RESPONSE MESSAGES...........14

Notice that in this example the appendixes are not numbered; in some companies the style guide requires that you number them. This sort of detail is just one of almost countless small differences among company style guides.

List of Figures and List of Tables

These are separate tables of contents for all the figures and all the tables in an analysis report. Each is made up of the figure or table number, figure or table title, and page number on which the figure or table appears.

List of Symbols

This is an optional addition to the front matter of an analysis report; include it if you think some readers will need to have symbols defined. The same thing applies to the inclusion of a glossary.

Introduction

The three-part purpose statement introduction appears here. It will orient readers to the main issue of the report, to the technical issues or specifics that are important to the report, and to what the report is intended to accomplish for the readers.

Discussion

The discussion contains an analysis of the technical issues important to the report. It supports the main issue of the report by providing evidence and explanations. It should be subdivided into topics, each with a subheading.

Conclusions

This section presents the results of the analysis, the evaluation of what was presented in the discussion. Sometimes listing the conclusions is a good, concise way to organize them. It calls attention to the conclusions individually, but still enables writers to explain them as is necessary.

Recommendations

Not all analysis reports have recommendations, but if they do, the recommendations tell the reader what to do with the information provided in the report.

Appendixes

The appendixes are not a dumping ground for anything left over. They should serve a precise purpose. For example, if you have information that is not vital to the understanding of the main and technical issues of an analysis report, but that is still important to groups of readers, put it in an appendix. Such information often includes derivations of equations, tables of raw data, sample code, and so forth. But the only way to be certain that what is placed in the appendix belongs there is to assess it within the context of audience needs.

CONCLUSION

In this chapter, we have briefly examined analysis reports. The reason the examination was brief is not that these reports are comparatively unimportant.

They are vitally important to companies. However, most of what was presented in the chapters about discussions and conclusions to articles can be applied to analysis reports. Most articles published in peer-reviewed scientific journals and in technological trade journals are analytical.

SUGGESTED READINGS

Hering, Lutz, and Heike Hering. *How to Write Technical Reports.* New York: Springer, 2010.

Riordan, Daniel G. *Technical Report Writing Today,* 10th ed. Boston: Wadsworth, 2013.

CHAPTER 17

How to Write Product Descriptions

Descriptive techniques are primarily used for writing product descriptions. But product descriptions are not only used to describe a product, they also are used to sell the product or to attract a reader's attention to it. Many companies use product descriptions as a part of their integrated marketing communications. In large companies, these documents are often written by specialists, but in small companies, you might be the specialist (more about this in later chapters).

AUDIENCE NEEDS

The first thing to consider is what the audience needs from a product description. Use the following checklist as a guideline.

1. What is the product? Notice that this is a definition question.
2. What is the product used for?
3. What does the product do?
4. How does the product do it?
5. What happens after the product does it? One of the sources of failure in a product description is when the writer does not realize the reader's need to know when the description is complete. Answering this question meets that need.
6. What is the product made of?
7. What are the product's basic parts?
8. How are the parts related to make the product do what it does?

You'll notice that the last three questions are those that you are likely to think of first. I listed them last for precisely that reason. For those questions to be answered effectively in product descriptions, you must first answer questions 1–5.

For example, consider questions 2 and 3. These questions focus you on the product's purpose. All products have a purpose. Communicating that to readers aids them in understanding the product itself. Does your product have a single purpose or many possible uses? Notice here that I will equate purpose with use; they are similar, and in product descriptions, they are one of the most important details. Does the product operate independently or in conjunction with something else? Who uses it? And under what conditions? A former student of mine at MIT wrote a brilliant product description of a welding device he and a classmate had designed. The description was well organized, and it read well. But it did not explain who was supposed to use the product and under what conditions. I later found out that this particular welding device was intended to be used by remote control under the ocean up to depths of 3,000 feet. It would attach bolts to pieces of sunken ships so that cable could be looped around the bolts and the ship salvaged. One could certainly argue (and I'm sure the U.S. Patent Office would) that this omitted detail was the most important aspect of the welding device. Don't overlook these sorts of details in your own product descriptions.

Treating issues of this sort makes your description more successful when you get around to describing the product's size, shape, and dimensions. When you begin to describe these physical characteristics, take care to orient the product in the reader's eye. Obviously, a photograph or line art will do a good job of this orientation, but not all product descriptions have that luxury due to budget restrictions. Realize that all products have a natural orientation, a side or view that a user would approach first. You will want to describe the product from that point of view. For example, how do you see a chair in your mind? I would bet you see the front of it, facing you, inviting you to sit down. That would be the view I would use to describe a chair. Think of how difficult a time an audience, which is not familiar with chairs, would have if you described it upside-down, or with the back facing the user.

Finally, don't forget the product's color and finish—if they are important aspects of the product itself. Color is important for fire trucks; it probably isn't for hard drives. Texture is important for nonskid surfaces, such as accelerator pedals; it probably isn't for virtual reality (VR) headsets.

PRODUCT DESCRIPTION PRINCIPLES

For product descriptions to succeed, you must keep your purpose and intended audience clear always. These two issues govern the extent of details used in the description, the kind of details used, and the order in which those details are used.

Beyond that, product descriptions also use three types of details:

- those that describe the product's function or use
- those that describe the product's physical characteristics
- those that describe the product's parts or components

Finally, product descriptions can be general (for unsophisticated audiences who are unfamiliar with the class of products) or they can be specific (for sophisticated audiences who already know what the line of product is). You are more likely to encounter the first type of audience when a product description is part of a longer document, an owner's manual for example. The second type of audience is most often reached by stand-alone product descriptions used in marketing and advertising communications. Each type of description has its own format.

FORMAT FOR A GENERAL PRODUCT DESCRIPTION

Begin by identifying the product, usually with a definition. Explain why the product is important to your readers. Forecast where your emphasis will be—on function, physical characteristics, parts, or some combination of each. Think of this part of a product description as your introduction. In the body of the description, describe the functions, physical characteristics, and parts. Any order of these details will work, but make sure it matches your purpose for the description.

If you begin the body with a description of the product's function, be sure that you focus on who uses the product and when, where, and how they use it. If the product is used in conjunction with some larger system, don't forget to describe the functional relationship with that system.

When describing physical characteristics, try to enable your readers to "see" the product. Focus on characteristics that appeal to the readers' five senses, obviously choosing those that are important to the product itself. The aroma of natural gas is important when describing a stove, but the aroma of a microchip probably isn't.

When describing the parts of a product, first list them and then describe them in the order of that list. For each part, you will want to define it, describe its function(s), its physical characteristics, its relationship to other parts, and if necessary, subdivide it and run through the same sequence of material for all the subcomponents.

Notice what is occurring here. I am suggesting that for general product descriptions you use a format that replicates itself. It is a nested format. For complicated products, you simply keep subdividing as often as necessary, repeating the description format at each level. Theoretically, I suppose one could eventually arrive at the subatomic level in a product description. Fortunately, I've never seen that done. But it may simply be that I have not had experiences with companies that are developing biological computer interfaces or quantum computing systems or nanotechnology. With these products, this approach to product descriptions will still work—even at the subatomic level.

General product descriptions close by showing how the components work together to make the product do what it does or by mentioning unique features of the product or variations of it.

FORMAT FOR A SPECIFIC PRODUCT DESCRIPTION

One might think that for a specific product description the format would be even more involved and complex than for a general description. But that is not the case. You can safely assume that the audience of specific product descriptions shares more information with you than the audience of general product descriptions. As a result, specific product descriptions are shorter, tighter versions of general product descriptions.

For example, in the introduction to a specific product description you would want to combine product definition and function descriptions. The body of a specific product description is limited to a description of the product's features.

Close a specific product description with any of the following:

- the last feature of the product
- a summary of the product's benefits and uses
- a description of product variations

For one of the best models of specific product descriptions (and a model that is easy to obtain), go to your nearest automobile dealership and obtain a new-car brochure. You'll notice that the cover describes, usually with a photograph, what the product is and the glamorous ways it can be used. The inside of the brochure describes the features of the product, and the back page often lists all the options you can get. Obviously, new-car brochures are more glitzy than most technological product descriptions, but that generalization does not hold across the board. Some new high-tech products are introduced with product descriptions that are every bit as flashy as a new-car brochure, as anyone who has been to trade shows can attest. Naturally, such descriptions are the result of large budgets. If your budget is not that flexible, simpler product descriptions work fine. Excellence is in the design and writing of the communication, not necessarily in its surface attractiveness or the size of the printing budget, as the following example (Figure 17.1) demonstrates.

Product Description: Ace Staple Remover

The Ace Staple Remover (Patent 4674727) is a US-made office product designed to safely and successfully remove staples from a variety of hard-copy paper documents. The staple remover is approximately 5.5 centimeters in length by 3.0 centimeters in width, at its widest point, by 4.0 centimeters in height. It is comprised of two "arms" connected by a pop rivet axle hinge that is spring-controlled and pressure activated, located approximately 0.75 centimeters from the small end of the staple remover.

The two arms of the staple remover are similar in design. Each arm consists of a steel rectilinear frame that tapers to two teeth located at the large end of the staple remover. The upper arm is 1.0 centimeter in width; the lower arm is approximately 0.7 centimeters in width. The different widths of the arms permit the teeth of the lower arm to interlock tightly with the teeth of the upper arm in order to securely grip a staple that is to be removed. The rectilinear frame is open to the inside of the staple remover to decrease weight and increase strength of the product. Viewed from either end, the pop rivet axle is clearly visible, with the spring coiled around it. This functional detail is designed to permit the staple remover to "spring" open after it has been used. On the outside of each steel frame, a plastic plate has been pop riveted to the frame in order to create a comfortable surface for ease of use. The plastic plate gradually widens from 1.0 centimeter in width at the small end of the staple remover to 1.5 centimeters in width approximately 2.75 centimeters from the small end. At this point, the plastic plate flares sharply outward to a width of 3.0 centimeters, so that a user can easily fit two fingers or a thumb onto each side of the staple remover in order to operate it. This section of the plastic plate is cross-hatched to create an easy-to-grip texture for the user. At approximately .5 centimeters from the long end of the staple remover, the plastic plate flares abruptly back to 1.0 centimeter in width.

To use the Ace Staple Remover, place two fingers on one flared plastic plate and the thumb of the same hand on the other plastic plate. Gently insert the teeth of each arm under a staple you wish to remove and apply smooth pressure in order to close the teeth and grip the staple. Then, with a gentle wiggling movement of the hand, remove the staple. Staples may be removed from either side of a document, but removing the staple from its smooth side can result in tearing the page (or pages) of the document; to avoid this, follow the procedure just described on the clasped side of the staple, rather than the smooth. This will cause the ends of the staple to bend outward to an open position. Then, repeat the procedure on the smooth side of the staple to completely remove the staple without damage to the document.

Figure 17.1. Product Description.

PART V

How to Write and Design for Digital Media

CHAPTER 18

How to Use the Internet in Professional Environments

Over the past 10 years, this topic has been the most popular request for seminars in my consulting practice. The Internet has made the global marketplace a reality and online communications the coin of that realm. As executive editor of the *Journal of Technical Writing and Communication* (Sage Publications) and the coeditor of *The Routledge Series in Technical Communication, Rhetoric and Culture* (Taylor & Francis), I communicate with authors around the world on an almost-daily basis. I receive manuscripts from authors as Word attachments or pdf files. I download and print them (my nod to old-fashioned, hands-on editing and to the fact that research has shown editing on the display screen to be less rigorous and less accurate than editing on hard copy). I edit them, marking up the copy as I go along. Next, I make the necessary changes to the online version, and e-mail the attachment back to the author with my suggestions for revisions. The entire process takes a little less than two hours, with the bulk of the time spent marking up the printed copy. It is hard for me to fathom the inconvenience of using the U.S. Postal Service for any of this (as many of us in fact did early in our careers).

That exchange, which is common to professionals who use the Internet as a daily professional communication tool, makes the entire process more productive, and in publishing shortens lead times between submission of manuscripts to appearance in print by weeks, if not months. But the speed and efficiency of Internet communications, along with its potential for long-distance collaborative work, have also forced us to rethink the processes and products of writing in that environment. Even today in the still-early 21st century, the Internet remains an immediate and relatively informal neighborhood with one substantial oddity—most of us will never meet face-to-face the people with whom we regularly communicate.

Virtual communications environments (e-mail, Skype, and other tools) alter our ideas about how we work together by imposing different views of communications context, particularly those involving time and space. Sending an e-mail document to recipients who are many time zones removed from us (as I do to the Sage production center in Delhi) transforms how we work, but it also means professional colleagues can participate in such collaborative efforts at a fraction of the cost. Moreover, the conversational nature of electronic mail makes the communication experience more personal than printed documents sent back and forth through more traditional means. Skype even permits face-to-face communications, reducing the necessity and expense of travel.

As Internet communications continue to improve, we must plan for enhancing their professional nature in our daily communications activities. This chapter and those that follow it in this section of the book will explore the peculiarities inherent in a variety of online communications, with an eye particularly toward professionalism in their use.

COLLABORATION AND THE INTERNET

The Internet's conversational nature holds great potential for the collaborative development of documents about technology. Teams involved in designing and writing a wide range of documents can be separated by almost limitless distances yet still work together to produce an organization's information products. Users can provide instantaneous feedback during beta tests. But explorations of these tools suggest the potential is not without some costs.

For one thing, writing on the Internet still and too often resembles typed conversations, so the potential for confusion that exists in any face-to-face conversation exists on the Internet, but it is exacerbated by the fact that visual facial cues and voice inflections are missing. Experience suggests that the possibility of misunderstanding and conflict, as well as confusion, can occur more readily than during face-to-face conversations. In the give-and-take environment of meetings dedicated to the development of documents about technology, conflicts also often arise. But many feel that it is the working through these conflicts that often leads to better products, as competing ideas and values are brought to bear on the development process. This creative disharmony—though at times painful to participants—is a valuable aspect of collaboration.

So if you plan to use the Internet as a collaborative document development tool, consider these points:

- Write as clearly as you would in a printed document. Don't be fooled by the conversational nature of this medium; confusion can occur easily. In fact, much of my seminar material about e-mail in professional environments is aimed at teaching people how to be a bit more formal when using e-mail at work.

- Be aware of cultural differences and the effects of idiomatic language when communicating to persons in other countries. The Internet can enhance the problems we occasionally experience in other written forms of documentation.
- PROOFREAD! Realize that hastily typed, poorly thought-out messages sent through electronic mail make a public relations statement about you, your qualifications, the company you work for, and the time, effort, and seriousness you devote to your work. Don't make bad first impressions. This continues to be a major issue at many corporations.

PROFESSIONAL MESSAGES FOR INTERNET COMMUNICATIONS

When we consider the number of daily, relatively informal messages written by professionals using e-mail during their careers, it seems wise to simplify the discussion of techniques used to organize and write these messages. One way of simplifying is to classify them according to the purpose they serve:

1. recommendation messages (either requested or self-initiated)
2. progress messages
3. informative messages
4. information-requesting messages

The design of each type of message varies slightly (in the case of progress messages, greatly), depending on these classifications. The rest of this chapter will be devoted to discussing these message designs.

Recommendation Messages

A recommendation message is used when writers wish to make a recommendation to the reader concerning some problem or issue that the writer has investigated. It might help to think of recommendation messages as a type of informal proposals. As was suggested above, the information contained in these messages can be either requested by the reader or initiated by the writer. If the potential reader has requested an investigation that will lead to conclusions and recommendations, the message generally adheres to the following design:

Introduction (reader-centered)—The introduction summarizes the problem as the writer understands it and provides a brief statement of the recommended solution.
Discussion—The discussion or body of the message presents an evaluation of all the alternatives the writer considered, including the pros and cons of each.
Conclusions and recommendations—This section recommends a solution, analyzes it, and discusses important aspects of implementing it.

And yes, e-mail software now provides writers with easy-to-use design tools, so that headings can be included in the message. Use them. The result will be e-mails that are quicker to read and easier to understand. And what reader would not appreciate that?

If, on the other hand, the writer initiates a recommendation message, feeling that the intended audience has a need for the information presented, it generally adheres to the following design.

Introduction—A three-part purpose statement introduction that supplies background orientation to the issue being reported on, a definition of the specific problem at hand, and a brief statement of the recommended solution.
Discussion—An evaluation of all the alternatives the writer considered, including pros and cons of each.
Conclusions and Recommendations—A recommendation of a solution, a full analysis of it, and the presentation of important aspects of implementing it.

Progress Messages

Progress messages (often referred to as progress reports or, less formally, as updates) are issued at predetermined intervals during an ongoing project. These intervals vary depending on the needs of the company. They may be daily, weekly, monthly, or quarterly. Regardless of the intervals, progress messages are used for one purpose: to report the progress made on a project during the interval covered by the message. They can do this most concisely when they adhere to the following design.

Introduction—An orientation to the status of the project by describing its background, including what has been accomplished prior to the present.
Discussion—An analysis of the accomplishments and problems during the interval being reported on.

Future Work—A statement of the work planned for the next interval.

Informative Messages

These messages are issued at the discretion of the writer to provide readers with information they "need to know." They are usually short, to the point, and often informal. Even so, for them to work, they should be well organized. The following design provides that organization.

Introduction—A three-part purpose statement that orients the reader to the subject of the message, provides a brief description of the contents of the message, and suggests the action the reader should take with regard to the information presented in it.
Discussion—The presentation of the information.
Requests—An optional section used if the writer wishes information from the reader pertaining to the contents of the message.

Information-Requesting Messages

These messages are issued when the writer has identified a need for specific information. Usually, they too are short. The following design makes it likely that the request will be noticed easily by the reader.

Introduction—An orientation to the writer's needs, including a reason why the message was written and why it was sent to the reader.
Discussion—A direct request for the information needed.

CONCLUSION

Information acquisition, storage, distribution, and presentation came together on the Internet, and literally changed communication in ways that are as significant as Gutenberg's invention of the printing press.

The Internet (and other digital technologies) has come as close as anything to date at creating a reality of McLuhan's global village; interestingly, it has reincarnated a modern form of letter writing among professionals, and even personal friends, which the telephone was predicted to eliminate. These letters, however, are not the finely wrought pieces of writing we associate with Victorian correspondence. Rather, they are more akin to one-sided conversations, typed on the fly, replete with error and poorly thought-out ideas. As

Frederick Williams, University of Texas communications professor, pointed out: The Internet can be used to produce "massive amounts of information, but little usable knowledge" (Freedom Forum Media Studies Center Leadership Institute, June 22, 1994). Nearly 25 years later, this remains too often an apt description. It is up to us who use the technology daily to ensure that we are not contributing to this type of information overload.

SUGGESTED READINGS

Felder, Lynne. *Writing for the Web: Creating Compelling Web Content Using Words, Pictures, and Sound.* Berkeley, CA: New Riders, 2011.

McCoy, Julia. *So You Think You Can Write? The Definitive Guide to Successful Online Writing.* CreateSpace, 2016.

Sammons, Martha C. *Longman Guide to Style and Writing on the Internet*, 2nd ed. New York: Longman, 2007.

CHAPTER 19

How to Design and Write for Multimedia Applications

Anyone who writes about technology sooner or later will confront multimedia. At its simplest, this task might involve designing the online help facility for a new product. It might involve developing online tutorials. But at its richest, it involves coordinating sound, video, photography, graphics, text, and interactivity to best communicate complex messages. A new and exciting development is the game industry

One thing to remember about this form of communication is that multimedia is not new; only the delivery technology is. If 25,000-year-old cave paintings were "used" in combination with song, dance, and mime for a tribe's instructional purposes, this event was multimedia. Also multimedia were medieval morality plays, which were developed in tableaux on a series of wagons that were pulled around a centrally located, stationary audience to depict and educate regarding the seven deadly sins.

But modern multimedia, as we think of it in the 21st century, is relatively recent. It originated at the Massachusetts Institute of Technology in the 1940s as an idea, but without the technology to make it happen. The idea was refined in the 1960s, but still technological limitations relegated it to a fantastic dream of universally located kiosks at which users could access an endless variety of information. Today that dream exists as the World Wide Web, and miniaturization made the dreams of Steve Nelson possible in ways he never imagined—smart phones instead of kiosks.

Fully operational hypertext systems were also first developed in the late 1960s, but it took mainframe computers to control the newly developed nonlinear architecture for making links in information and providing access to it. The Aspen Movie Map, created in 1978, was the first promising development. In this experiment, researchers drove a flatbed truck through Aspen, Colorado, taking movies as they proceeded. At each intersection, they filmed in all four directions; they even filmed going into businesses. The result, once everything was loaded onto a computer, was the first true multimedia tool coordinating motion visuals, sound, and interactive user control. Today, Google maps makes such technology commonplace.

This chapter will explore how to design and write for multimedia applications. It will not examine multimedia development programs, since they change, update, and are replaced rapidly. Rather, the goal is to provide a structured design process to ensure, as has been done with written documents in each of this book's editions, that you develop multimedia products that work for users.

A DESIGN PROCESS

Creating any multimedia product requires a structured approach to the design process. Multimedia is complex and multifaceted; it is rare that any one product developer is expert enough in all its facets to work alone. This need for help means that all of the personality-associated complexities of entrepreneurial teamwork come into play. Managing the team helps manage the design.

Building a Design Team

In creating a multimedia design team, you not only want to select people who work well together and who deal efficiently with mountains of complexity, but you also want a variety of skills brought to the design. At its fullest, a multimedia design team should include people who have knowledge of the following:

- audio
- video
- digital photography
- graphic design
- technical documentation
- training, learning and game theory
- character and plot development
- authoring languages, including virtual reality
- quality assurance testers
- project management

What we have seen over the past 20 years is that multimedia development, particularly as it is applied in the game industry, has become more specialized, more complex, more demanding.

Creating a Design

Creating a multimedia design is a four-phase process:

- developing the design
- developing the documentation
- developing the media
- authoring

These phases are not necessarily sequential in the team environment of multimedia development. Different subgroups within the design team can work simultaneously on developing design overviews, media, and documentation; and the more complex the project, the more likely this will be necessary. In entrepreneurial design teams, the process is often iterative and experimental.

Developing the design begins with determining what the customer wants. In other words, what is the multimedia product supposed to do and for whom will it do it? Will it be a stand-alone kiosk in a trade-show display? A training tool? A web page? A help facility? An engaging game? Time should be spent early in the design process to nail down these issues to avoid considerable problems later when more specific product elements are being developed.

Multimedia is essentially a script medium early in the development phase. The design team should create treatments for each independent segment or module that will be incorporated in the final product. A treatment is essentially a short narrative. Consider the following example:

> *Kids Tour Boston Treatment*
>
> The Kids Tour Boston Treatment provides a prototype design for a tablet-based multimedia children's tour that, when fully implemented, may be applied to any tourist location in the city. From the opening sequences to the categories at the lowest hierarchies, all aspects of the program can eventually be applied to any city.
>
> *Opening*
>
> The opening sequences are designed to be intriguing and attention getting. They also orient the user geographically. Our prototype zooms in on Boston, Massachusetts; subsequent productions will utilize the same format: Earth, continent, country, region/state, specific target location.
>
> Beginning with a shot from space, the screen will zoom in on the spinning planet Earth. The globe slows and finally stops on the North American continent. North America then changes from green to a pulsating yellow. It rises off the Earth and fills the screen.

In North America, the United States is outlined and turns pulsating red. It lifts off North America and expands to fill the screen. In the United States, Massachusetts becomes a pulsating blue. It then lifts off the United States to fill the screen. In Massachusetts, a black blinking dot marks Boston.

The black dot rises quickly, turns into a square and expands to fill the screen and then resolves into a panoramic view of Boston.

As this sequence, a narrator (child's voice) says: "Welcome to Boston, the capital of Massachusetts."

As soon as the voiceover narration concludes, the words, "KIDS TOUR BOSTON!" fade into the screen in bright yellow lettering. The screen background turns black and the title remains slightly above the center of the screen.

The mascot/tour guide appears: the Bean. This is an anthropomorphized Boston baked bean. He is a small reddish-brown cartoon character with skinny legs and arms. He has big round eyes and a wide, friendly grin.

(The treatment continues to specify details of the tour program design.)

Once treatments have been developed, the design team should resolve the following associated issues:

- **Media selection.** Begin determining the audio, video, and still images that will be included in the product.
- **Motivation.** Assess how the product encourages customers to use it. The higher the motivation, the longer (and more rewarding) the use. Multimedia and game developers call this "seat time," the period of time the customers stays engaged (or in their seat) with the product.
- **Interactivity.** Just how will the user interact with the material in a multimedia product? Will it be through a touch screen? By typing? Voice recognition? Joy stick? VR headset? How much interactivity will be available? Remember: the more interactivity, the higher the motivation; but also, the greater the complexity. Each of us have no doubt been "lost" in a game or a website.
- **Sequence and structure of information.** How information is organized in a multimedia product is crucial to its success. Multimedia makes use of asynchronous organization and takes advantage of the human proclivity to think associatively rather than linearly. It is this nonlinear organization

that accounts for the richness of multimedia. Flow diagrams can be an important primary organizational tool for this effort.
- **User evaluation.** In any product that requires customers to interact with it, it is important to develop ways that the users can determine whether they are being successful. These can be simple audio cues or visual reinforcements that correct decisions are being made and that the product is responding appropriately.

Documenting multimedia product design and development produces mountains of information. This information is crucial for developing media elements, for the authoring stage, and for the development of product literature ranging from user documents to marketing and advertising materials. Documentation includes at least the following elements:

- Reports explaining design decisions and rationales balanced against customer wants and expectations. These are analysis reports, and the information they contain lives on in other types of product documentation.
- Video storyboards depicting every visual scene incorporated in the product, as well as the activities contained there.
- Audio scripts for narration and dialog that are included in the product.
- Shot lists for all photography to be included.
- Art rendition lists for all artwork.
- Graphic rendition lists for all computer-generated graphics.

Developing media elements is a multidimensional aspect of multimedia design all to itself. Creating audio and video clips from scratch might involve producers, directors, actors, production crew, and editors. Sites have to be chosen or scenes created. Narration and dialog have to be developed and practiced. Special effects may be developed. The action gets filmed. All the versions are edited to a final copy. Animation is similarly complex, as well.

It might look simple when you are playing a mobile game, but the work in developing media elements for a multimedia product is immense . . . and time-consuming. For example, my university wanted a short segment of one of my classes for promotional use on the university website. This shoot was to be about as simple as one could imagine. The actors were set; they were my students. My role involved walking from the back of the class to the front, as if I were in the process of teaching. Absolutely no audio would be used. The shoot took four hours—not including the production crew's time to set up and take down equipment in the classroom. The length of the final segment? Twenty seconds! This very complexity has led to the development of stock-footage audio and video libraries, which developers now regularly use.

Finally, once all the design and its myriad associated components are complete (and even during the design process, in some cases), the project is turned over to multimedia authors, specialized programmers who use authoring software to create the product's final structure, incorporating the necessary links so that users can access desired information. Depending upon the size of the organization, this work might be contracted out, but as authoring software has become simpler to use, more developers are bringing this task in-house.

CONCLUSION

In this chapter, we have examined a brief and very simplified overview of multimedia design. The subject is complex, and entire books are available to explain it in detail. Regardless of the specifics, multimedia is but the latest step in developing usable communications about technology. Every guideline that applies to the shortest e-mail or to the clearest proposal applies to multimedia products. But multimedia is not text; it is more than text. Developers must understand the conceptual differences and try to limit the use of text as much as possible. Keep sentences shorter than you would for hard-copy text. Use the command voice and active voice constructions. Locate small "chunks" of text strategically in various places on the display instead of grouping all text together in one location (people won't read it). And remember, multimedia communications must be clear and easy for users to understand. Pay particularly close attention to the guidelines that follow to conclude this chapter.

DESIGN RULES OF THUMB FOR WRITING FOR THE WEB

Screen Design

- It takes 4–6 display screens to effectively communicate the information contained in a single printed page.
- Display screens are typically arranged in a landscape format (wider than tall); printed pages are arranged in portrait format (taller than wide); the difference affects how well users can gain information from a display screen. Don't crowd screens with text.
- Do not exceed 30 percent text density. The text on a screen should not occupy more than 30 percent of the overall display space.
- Write concisely, even more concisely than in print, but not to the point of using a telegraphic style. Online information should still sound like natural language to readers.

- Use tabular formats. Lists, boxes, tables, and other display strategies help manage the bulk of information by organizing it into easily scanned groupings.
- Keep lines 12–14 characters long. Human eyes fatigue more quickly reading from a display screen; you can help alleviate that condition somewhat by avoiding long lines and using multiple columns.
- Keep paragraphs 6–7 lines long. Same issue here. Make information quick to scan.
- Standardize window placement and organization. Companies and individuals should try to standardize their use of window placement and organization within online products. Have pop-up windows appear in the same general location each time you use them, so that users become conditioned to where new information will appear. Be consistent in how information is organized within windows. Random order and random organization conspire to confuse users.
- Present related information vertically in lists; vertical lists are easier to scan than horizontal arrangements of the same material.
- Use indentation to show subordinate relationships.
- Use labels, headings, bullets, dashes, and other visual cues to help readers locate information on a display.
- Be careful with fonts, justification, leading (interline space), inter-paragraph spacing, and the placement of visuals. Some fonts work better than others for online displays. Left justify text and leave right sides ragged; while the jury is still out on this particular issue, the preponderance of evidence suggests that right-ragged text is easier to read than right-justified text. Use more leading for online displays than you would for paper-based text; the increased space eases reading, as well as eyestrain, and promotes better scanning of the information. The same is true for inter-paragraph spacing. Put visuals where the users will notice them.
- Use the Gutenberg Diagonal. This concept has to do with how a reader's eye scans a page or a display screen in an elongated Z from top left to bottom right (Figure 19.1). Place important information (both text and visual) along this line.
- Make each individual screen as self-sufficient as possible. Needless jumping back and forth among related screens decreases comprehension of the information and overall usability of the product.

Links

Links show the relationships in information by interconnecting topics. They are what we use when we provide access to separate topics in all online

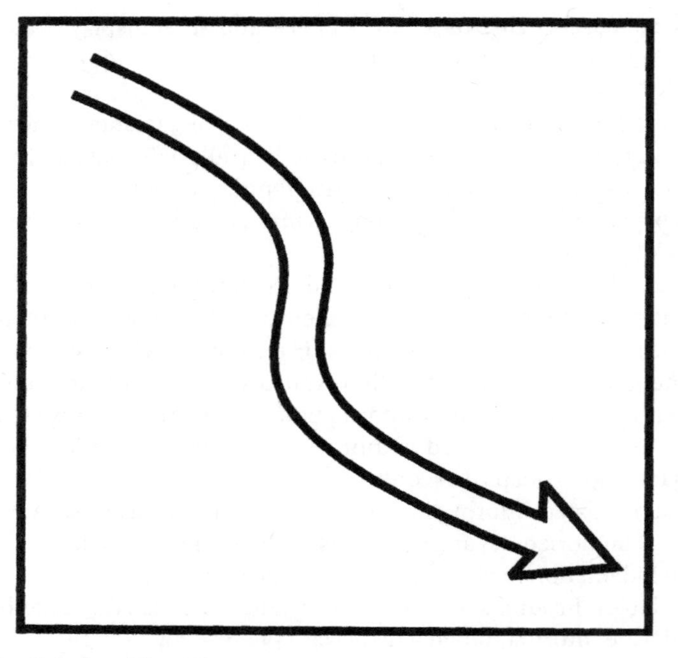

Figure 19.1. Gutenberg Diagonal.

information systems from web pages to online tutorials or when we provide access to the same topic from different points located throughout a multimedia product, or guide readers to related topics. Links establish the associational organization inherent in multimedia products.

Links are activated by "hot areas" on a display screen—most often icons or highlighted text. Usually, when a mouse is dragged across a hot button, it identifies itself in some way—most often by becoming highlighted or by changing color or blinking. Clicking on the interactive area activates the link contained there and takes the user to a new piece of information. Links can either provide new information, or they can cause some media event to occur, such as running a video clip or activating an audio file or causing a connected piece of external equipment to run.

When designing links, take the following advice into account:

- Links should be predictive. It should be clear to users what is going to happen when a link is activated and how to undo it. Be as obvious as possible when indicating what a link does.

- Links should be worded clearly.
- Links can allow for multiple destinations, but the users' choices should always be clear.
- Links should not add clutter to the screen design. For example, a link should only become highlighted when the mouse is dragged over it. The more links on a single screen, the more the complexity, and the more difficult the page is to understand and use.
- Links should be easy to use. Basically, this means making the interactive area obvious. For mouse-activated links, the minimum size for a hot area should be about 1/4-inch square; the same size is true for trackpad systems common to laptops. For stylus systems, the minimum size can be reduced to a 1/5-inch square. For touch screens, it should be increased to a 1/2-inch square.

THINK BEFORE YOU LINK

Some sort of logic should be apparent to users in the design and inclusion of links in multimedia products. When considering whether to include a link, ask yourself whether a significant number of users will want to stop reading or doing whatever they are engaged in at the present moment in order to be transported to related information or tasks elsewhere. If the answer is "no," don't put a link there. Ask yourself where users will expect links to occur, or where they will want additional related information. And most importantly, ask yourself if users will understand what they find when they activate a link, as well as the reason for going there.

SUGGESTED READINGS

Garrand, Timothy. *Writing for Multimedia and the Web,* 3rd ed. Boston: Focal Press, 2006.

CHAPTER 20

How to Design and Write for Social Media

This chapter presents strategies for using social media (including e-mail) in professional environments. The strategies are intended as general guidelines, rather than specific rules for particular social media products, since those products are notoriously quick to change, cease to exist, or be replaced by newer developments. The strategies, however, are based on established rhetorical principles that are simply applied to these new developments.

REPORTS AND PROJECT MANAGEMENT

In the 21st century, much of the daily to-and-fro communication that is used to manage projects occurs through e-mail. Given the informal nature of e-mail through its early use, this can be a challenge in professional environments; so it is important to understand the flow of information through an organization while projects are ongoing. A chart will help (Figure 20.1).

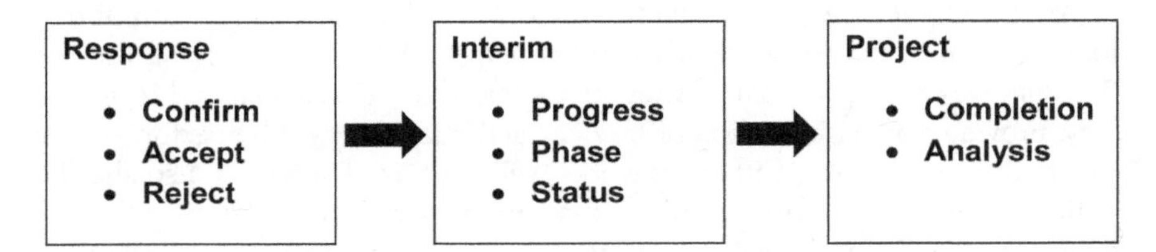

Figure 20.1. Flow Diagram of Project Management.

Projects are initiated by requests, which may be for information or action. If a request is for information, it can simply be answered in an ensuing e-mail, unless the recipient is unable or unqualified to do so. Then the request could be forwarded to someone else, which is a particularly helpful facility for the e-mail management of projects. I suspect that most of us do this daily; I do.

For requests that initiate action, a series of possible events can occur. The first is the response to the request. Ordinarily, we might think that such a response is binary—the recipient either says "yes, I can do that" or "no, I cannot do that." But in fact, there is a third option that needs to be considered first.

There are times when the recipient of a request cannot immediately determine whether or not he or she can do the request. The recipient needs more information, or more time to consider the request. No doubt, we have all experienced this situation . . . as requestors. We send a colleague an e-mail requesting that something occur. Time passes, with no response. An hour goes by, a few hours, perhaps even a day. By that time, we begin to wonder if the e-mail went through. We check our "Sent" file; everything seems fine. We wonder if the recipient missed it, ignored it, if the request wound up in the recipient's "Junk" file, etc. All of this wastes time and diminishes efficiency. But there is a simple solution—the Confirmation Response.

The Confirmation Response is a simple e-mail sent by the recipient back to the requestor—only in situations that might take a little time prior to writing a complete response. The Confirmation acknowledges receipt of the original request; it explains that additional time is needed to consider it (and gives reasons for this). It might request that the sender provide additional information. It might transfer the request to someone else who is more qualified to deal with it. And it forecasts how long it will take for a complete response. Notice what happens when Confirmation Responses are used: efficient two-way communication is established across an organization. Expectations are set. Time and worry are lessened. Consternation and wasted efforts are avoided. This can be the first step toward more productive in-house communications.

But most of the time, a recipient can say "yes" or "no" to a request. These situations still require a clear, well-designed response in order to create a productive communications environment.

An Acceptance Response (saying "yes") should include a minimum of the following. In it, the writer clearly accepts responsibility for satisfying the action that was requested. The writer should analyze the request, and if necessary, provide a specific outline of tasks or activities that will be performed to satisfy the request. A Gantt chart works well for this. The writer also should anticipate and clarify any problems that might occur in performing the request. Finally, the writer asks for a "go-ahead" if the requestor and responder are in agreement as to what will be done. Once again, this type of a response to a request for action delineates lines of communication in an organization; it enhances expectations and, inevitably, productivity.

There are also times, somewhat rare, when the recipient of a request has to say "no." These Rejection Responses are rare for obvious reasons—in most organizations the persons requesting other persons to do things are their superiors. Employees generally do not say "no" to superiors; doing so can be unhealthy to long-term employment. But there are times in a career when such a response might be necessary. These include requests that are illegal or that violate the recipients professional or personal standards of ethical behavior. As a result, we need to know how to do these responses in ways that are clear and that, as best as possible, can protect our employment. Successful Rejection Responses must clearly document the writer's viewpoints and conclusions; they must *prove* the negative response and the reasons for it. To do this, one must state opinions clearly and honestly; document reasons for rejecting the request. But as much as possible, recognize the requestor's needs, as well. Look for ways that a potentially bad request can be made workable. Propose alternatives. Reasonable alternatives, once again, promote two-way communication and increase the likelihood of success. They also help establish trust among coworkers in an organization.

Once a request has been agreed to, some sort of project begins—often based on the tasks listed in the recipient's Gantt chart (perhaps as amended by the original requestor). In most organizations, some sort of progress report is to be expected that will delineate what is occurring throughout the life of the project.

Progress reports detail the accomplishments on a project over a specific period of time. They might be issued daily, weekly, biweekly, monthly, quarterly, etc. The most effective progress reports examine and report on not only the current period being examined, but they also refer to unresolved problems from previous reporting periods, including what has been attempted in order to solve the problems. Effective progress reports also forecast the objectives of the next reporting period. As a result, a lot of information is collected and disseminated through such progress reports. Importantly, this information becomes source material for the analysis reports that conclude projects (see Chapter 16).

After a project is completed, typically some sort of project completion report (which I called analysis reports in Chapter 16) is written to become the official documentation of the project. These reports at a minimum analyze initial requests, compare the resulting project with those requests, and ask for approval at the conclusion of the project.

So far, we have examined a macro-level methodology for using e-mail and other social media to manage projects in professional environments. In the section that follows, we will consider strategies for the micro-level, the writing itself.

CHARACTERISTICS OF GOOD WRITING

Given the informal nature of social media use in our nonprofessional lives, it is important to understand the distinctions between that use and professional uses. Four categories are useful: Appearance, Writing Skill, Writer-Reader Relationships, and Rhetorical Strategies. I have listed these in the order that readers notice them.

Appearance

The first thing a reader notices in any communication is how it looks; specifically, if it looks inviting, easy, and quick to read. Writers, even of brief e-mails, should make sure that the message is designed so that it is clear and unobtrusive. Early in e-mail experience, this was a challenge; but in the 21st century e-mail software provides almost as much text design capabilities as print does. Use them. The one thing that must be avoided is long, unbroken passages of text. Readers simply will not read these.

Break up text passages as much as possible. Balance text, images, design elements, and white space—just as you would in a printed document. Does this take more time? Certainly, but the readability and accessibility gained is worth the time and effort. Your readers will appreciate you.

Writing Skill

Once readers have gotten past how your communication looks, the next thing they will notice is how well you have written it. Use position, spacing, diction, and syntax strategically. For example, put important information early in the communication. We know that readers tend not to finish reading e-mails; don't let vital information go unread, simply because you put it near the end. Think about ways to space information so that it is more accessible. These include bulleted lists and numbered lists, both easy to do now in current e-mail systems. Readers read e-mail quickly; use shorter, familiar words that promote this. Write in a conversational tone, even in professional environments. But understand the important differences between professional and nonprofessional uses (for example, there is no place *ever* appropriate for using the word "Hey" as a greeting in a professional e-mail). Conversational tone in professional e-mail communications is achieved by varying the length and type of your sentence; it is what we do naturally when we speak.

Write about people doing things. This automatically makes information accessible to the people who are reading it. And it virtually requires you to write in the active voice, rather than the passive voice—even if you have no idea how to distinguish the two.

Use concrete, specific terms rather than generalities. Provide plenty of details. In other words, realize that just because you are using e-mail as a communications medium, the standards for successful communication are much the same as for writing and printing a document.

Edit the results. That's right—edit your e-mails before you press "Send." You will thank me for doing it, because every one of us has had the experience when we were rushed, did not look over an e-mail after writing it, hit "Send," and immediately thought . . . "Wish I hadn't done that." So, check the spelling, check the grammar, check the usage, check the tone, etc.

Writer-Reader Relationships

Every time we write and send an e-mail, we create a relationship. And like all relationships, we should want it to be a good one. There are some things to avoid that will increase the likelihood of successful relationships.

Avoid condescension. In other words, never talk down to readers, even (particularly) in situations in which you are much better versed in the information you are communicated. The point of a successful professional communication is not to impress your readers with your erudition. It is to communicate, to enable something useful to happen. Talking to readers as if they were stupid is a guaranteed way of failing. The reason? Readers hold the power in these relationships—they can choose not to read.

Avoid anger. It never works in e-mail communications. Once again, readers will simply decide that you are not worth their time.

Avoid accusations. Written accusations, libel, and defamation can get writers in a lot of trouble quickly. Realize that e-mail (whether you delete it or not) lives forever somewhere, and it is "discoverable"—the legal term for lawyers finding it and using it in lawsuits. If you would not print something on top of the fold in the *New York Times,* leave it out of e-mails. The most recent presidential election proved fairly conclusively the damage that can occur when e-mails (supposedly private) are hacked and provided to an organization such as WikiLeaks to publish to the world.

Avoid sarcasm and cynicism for the simple reason that most of us do not know how to do this well. Our readers will think that we are being serious, and resolving that confusion is time wasted. Best not to attempt these things at all.

Rhetorical Strategies

Basically, these are techniques for being persuasive—even in communications in which you are not trying to sell someone something, because in a sense, you are. When we are writing the most analytical or objective of communications, we are still trying to "persuade" our readers that our analyses are accurate, based on well-chosen facts, etc. From a rhetorical perspective, all writing is persuasive.

With that in mind, organize information—even in a brief e-mail—according to your purpose. What this means is, as I have already said, put important material early in the communication, and support it with clear examples. Write from the perspective of your readers (audience analysis, again). Be positive, and be "success conscious," a trendy term for your absolute conviction that if you apply what you have read in this chapter, your professional use of social media will be appropriate, and it will work.

CONCLUSION

In this chapter, we have examined a range of strategies that have withstood the test of time over decades of designing and writing documents in professional environments. What we have seen is that, with a little adjustment, virtually all of these strategies are applicable to the 21st-century use of digital communications in professional environments.

In the end, good writing is good writing.

SUGGESTED READINGS

Standage, Tom. *Writing on the Wall: Social Media—The First 2000 Years.* London: Bloomsbury, 2014.

Tuggle, C.A., and Forrest Carr. *Broadcast News Handbook: Writing, Reporting and Producing in the Age of Social Media.* New York: McGraw-Hill, 2013.

PART VI

How to Write and Design Associated Communications

CHAPTER 21

How to Write Public Relations Documents

Virtually every organization from the smallest club to the largest corporation has a person, group, or division in charge of managing the public's perception of that organization. In small organizations, it might be a publicity chairperson who sends announcements of meetings to a local newspaper; in large corporations, it might be a public relations division that coordinates all aspects of the company's public affairs. Although you might think that publicity and public relations are the same, there is a key distinction between them. Publicity is nothing more than making announcements to inform a public about the operations of an organization. Public relations is managed communications; with it, organizations attempt to manage public perception regarding the organization.

Managed communications of this sort lead to a snake pit of problems, because public relations, as a profession, has had a history of bad public relations—from political spin doctoring to corporate half-truths to barely concealed falsehoods. Therefore, the best place to begin a chapter on writing public relations documents is with a clear description of the ethical environment in which the task *should* be performed.

The Public Relations Society of America (PRSA) has established voluntary code for standards of truth, accuracy, good taste, fairness, and public responsibility, which can be found on their website (https://apps.prsa.org/AboutPRSA/Ethics/CodeEnglish/index.html).

This code of honor, if judiciously followed, will ensure that all public relations activities we might find ourselves involved in or contributing to are performed at the highest ethical levels. With that in mind, in this chapter we will examine strategies for creating and writing public relations documents. These strategies will be useful if you participate in the writing of such documents or if you only participate in generating the information your company uses in such documents.

PUBLIC RELATIONS PLANNING

Successful public relations documents are best used in an organized public relations campaign, and such campaigns begin with planning to determine the best uses of a wide range of media and messages. Consider the following questions and directions.

- What goal or goals are to be accomplished by the public relations communication?

  Think of the answer to this question as the purpose of the communication. It might be to inform the public in a town that a new building will not detract from the town's character; it might be to inform employees about a newly hired CEO; or it might be to explain recent corporate decisions. The list of possible purposes for public relations communications is almost inexhaustible, but each should have one point in common—clear, honest communication.
- Who is the target audience for the communication?

  Try to answer this question as accurately as possible. Is it the general public? The public within some geographical boundary? Employees? Certain employees?
- What is the audience concerned with or interested in with regard to the communication?

  This concern is really no different from our desire to know the end users and to meet their needs in any communication about technology. Public relations messages, like all professional communications, are targeted to specific audiences, their knowledge, and their needs.
- Create the message.

  Based on your answers to the first three questions, draft the content for your public relations communication. It is important to write this draft before focusing on formats, because those formats vary widely depending upon the medium you are writing for.
- Select the communications medium (or media).

  This issue determines the message's format. Are you best served by selecting a print medium, such as newspapers or trade journals? Or should you be writing for a broadcast medium, such as radio or television? In the 21st century, increasing amounts of public relations communications are distributed through social media, because the impact is immediate. President Trump has proven to be a skillful user of Twitter™

to control the mass media news cycles, regardless of what one might think of his politics or motivations, and if I were to make one prediction in this book it is that we will see more of this as organizations catch up to the power of instantaneous public relations.

- Select a spokesperson.

  Selecting a spokesperson for delivering a public relations message may seem a bit different from other types of communication, but in reality, it is not. All written documents have a voice inherent in them; in nonfiction communications about technology that voice is most often the writer's. In certain public relations documents, press releases for example, it might also be the writer's voice. But in others, such as public service announcements read on a radio broadcast, a different spokesperson might be chosen. It could be a company representative or a narrator hired for the quality of his or her voice. But in either case, the key issue in selecting a spokesperson for delivering a public relations message is trustworthiness.

- Finalize the message or document.

  Depending upon the medium chosen for delivering a public relations message, you now confront the issues of format. In the following sections of this chapter, we will examine the two most common formats for public relations messages—press releases and public service announcements.

PRESS RELEASES

Press releases are much more than announcements made to garner some publicity for an organization. They are targeted, persuasive communications designed to pique the interest of a newspaper or trade journal editor or a radio or television news director so that more in-depth coverage of an issue might be developed. The goal in designing and writing an effective press release is to obtain positive news coverage for an organization. As a result, the key component of any information communicated by a press release is that it be newsworthy. It must be about something that the audiences of the paper, journal, radio, or television will find important. In other words, thinking back to the planning questions described earlier in this chapter, the information in a press release must meet the eventual audience's needs, interests, and concerns. Before editors or news directors are willing to commit staff and resources to develop a story, they have to be convinced that the story is newsworthy to the people who read their papers and journals or who listen to the radio or who watch the television news program.

Once the issue of newsworthiness has been determined, writing the press release requires adhering to the following criteria.

- Make sure that the information is new, different, or unusual.
  Does the information have a sense of immediacy or timeliness? Is it about something the audience needs to know now? The larger the size of the audience affected, the greater the chance editors and news directors will want to develop it.
- Make sure that the writing is concise and quickly paced.
  Editors and news directors are gatekeepers; they determine what gets in and what stays out. Writing successful press releases means getting and holding their attention. To do this, use the active voice; it reads quicker. Avoid terms that are unfamiliar to the reader. Vary sentence lengths and types; this variation keeps your writing style interesting. And make sure that the writing is flawless, with no punctuation, spelling, or grammatical errors. Limit the length of a press release to one page (two at the very most).
- Make sure that the release's lead works.
  The lead in a press release is the first paragraph. It must attract the reader's attention by identifying what is important in the message and by promising newsworthy material in the body of the release.
- Make sure that the body of the release is clear and concise.
  The body of the press release is where you support or prove your contention that the information is newsworthy and should be developed further. Focus on brief, clear statements of fact that are of interest to the audience.

Figure 21.1 is an example of a press release. Notice the use of headings and the placement of information that identifies who is responsible and who should be contacted for further information. Press releases identify the date on which the information can be used, or if it is irrelevant, they simply state "For Immediate Release."

PUBLIC SERVICE ANNOUNCEMENTS

Public service announcements are the other commonly used public relations documents. They are announcements of public interest designed to be read over the radio or in public affairs programs on television. Part of the federal licensing requirements for radio and television stations stipulates that the stations provide time for such public service announcements.

Writing public service announcements is probably the most challenging writing any of us is likely to do. The reason is that the announcements are limited by extraordinarily short time allotments. Public service announcements usually run 60 seconds (a long one), 30 seconds (an average one), or 10 seconds (a very short one). The best way to understand the writing challenge

FOR IMMEDIATE RELEASE

Contact:
Jim Jones
Pet Boundary, Inc.
P.O. Box 555
Westborough, MA 01581
508-555-5555
Jim.Jones@petboundary.com

PET BOUNDARY, INC. POSTPONES PRODUCT LAUNCH

Westborough, MA. November 9, 2016. – Pet Boundary, Inc. announced the postponement of the PetBoundary® Manager launch. Due to a recent change in suppliers, the company has opted to delay its proposed October release to conduct further field-testing using the new components. A revised launch date for the product has yet to be determined.

The PetBoundary® Manager is an electronic household system that uses radio control and computer technology to create up to twelve boundaries and manage boundary strategies for up to four pets. The system offers such feature as variable level programming, four different correction settings, history reporting, and system alerts, and promises to take existing technology to a higher level, allowing users to be more proactive about boundary training.

Since its public unveiling in July, the PetBoundary® Manager has been very well received at veterinary trade shows and Pet Boundary, Inc. has received hundreds of inquiries about the product. According to Pet Boundary's president, Susan Smith, the postponement of the launch was a necessary risk the company chose to take. "The delay is certainly regrettable, but we would rather ensure the product exceeded the expectations set for it, than release the product without a thorough performance check."

The PetBoundary® Manager was developed by JTC International and will be sold and distributed exclusively through Pet Boundary, Inc. via their website. A basic system will sell for $659. More information about the system can be obtained at www.petboundary.com.

--###--

Figure 21.1. Press Release.

involved in developing public service announcements is to convert these time limits into words and lines.

Based on a 60-character line read at average speed, a 60-second public service announcement is composed of 16 lines of text. A 30-second announcement has 8 lines of text to work with, and a 10-second announcement allows for approximately 2–3 lines of text. This brevity means that words have to be chosen very carefully; you don't have the luxury to waste a single syllable. Writing successful public service announcements requires constant revision to determine if another phrase, word, or syllable can be cut or condensed while still maintaining the message's integrity.

Typically, the time limits placed on public service announcements mean that writers can only focus on a single, important idea—introducing it in the brief lead statement and developing it in the brief body section that follows. The same criteria identified for press releases apply to public service announcements: they have to present newsworthy information; they must address an audience's interests; and they must be concisely written. Figure 21.2 is an example of a public service announcement.

PUBLIC SERVICE ANNOUNCEMENT

Time: 30 seconds

ALONE AND LONELINESS—THERE IS A DIFFERENCE

AND THAT'S THE THEME AT STATE UNIVERSITY THIS SEPTEMBER AS THE COLLEGE UNVEILS AMERICULTURE—A MONTH-LONG FESTIVAL THAT BRINGS TOGETHER A VARIETY OF VISUAL AND PERFORMING ARTS.

THEATER, MUSIC, PHOTOGRAPHY, DANCE, POETRY, COMEDY, PUPPETRY.

FEATURING STUDENTS AND FACULTY FROM STATE UNIVERSITY, HIGH SCHOOL STUDENTS, AREA RESIDENTS AND VISITING PROFESSIONALS. DOZENS OF EXCITING PERFORMANCES FOR THE WHOLE FAMILY.

FOR A COMPLETE LINEUP OF EVENTS, VISIT OUR WEBSITE AT WWW.STATE.EDU. That's WWW.STATE.EDU.

STATE UNIVERSITY—THE LEADERSHIP COLLEGE

Figure 21.2. Public Service Announcement.

SOCIAL MEDIA AND PUBLIC RELATIONS

There has been no change more revolutionary in the discipline of public relations than the adoption of social media platforms for public relations (PR) messages. E-mail blasts were the first example of this; essentially they are press releases and public service announcements sent to e-mail lists that PR professionals gain access to, often by purchasing the lists from other organizations. When e-mail blasts appear in one's e-mail inbox, it is impossible to not notice them, even if one quickly deletes them. Designed well, the headline—if nothing else—catches a reader's attention before the e-mail document can be deleted. And research shows that this might be just enough to plant an idea in the mind of the reader/viewer. Plus, if successful (and not immediately deleted), readers read these messages. In many ways, these are press releases and public service announces are sent directly to the public, without the historical filters of editors, journalists, program directors, etc.

Twitter™, as was mentioned earlier, seems especially well suited for immediate, brief PR messaging.

And the public relations subdiscipline, known as Crisis Communications, is ideally suited to social media environments. The word "crisis" implies the importance of immediate communications. Most university campuses, as well as other businesses, now have protocols for immediately notifying students and faculty about such crises as active shooters on campus, as well as severe weather events.

If you think of adopting public relations as part of your professional communications toolbox, be sure to include social media, and use it wisely.

CONCLUSION

In this chapter, we have taken a brief look at the development of public relations documents, focusing primarily on the two most common—press releases and public service announcements. As professionals employed in high-technology enterprises, we might never actually write or plan such documents. More and more companies contract these efforts out to public relations firms. But if you are an entrepreneur, or if you work for a small startup company, you most likely will not have such a luxury as hiring public relations professionals to do this work. You will have to do it, and to do it well, you need to know how to avoid common mistakes in creating successful press releases and public service announcements that establish the best public relations for your company and the people who work there. The suggestions in this chapter, including the importance of social media as a public relations delivery system, will help.

SUGGESTED READINGS

Newsom, Doug, and Jim Haynes. *Public Relations Writing: Strategies and Structures,* 11th ed. Belmont, CA: Wadsworth, 2016.

Wilcox, Dennis L., and Bryan H. Reber. *Public Relations: Strategies and Tactics,* 11th ed. Boston: Pearson, 2014.

Wilcox, Dennis L., and Bryan H. Reber. *Public Relations Writing and Media Techniques,* 7th ed. Boston: Pearson, 2012.

CHAPTER 22

How to Write Marketing and Advertising Documents

Just as was the case with writing public relations documents, you might think you will never find yourself engaged in writing information that will be used for marketing and advertising purposes. But it is more common than you might imagine. For example, the technical descriptions that are included in specifications and other reports eventually can find their way into product brochures and even advertisements placed in trade journals. The automobile brochures and booklets we pick up when car shopping or at trade shows demonstrate this very well.

Another possibility for writing marketing and advertising materials exists for the entrepreneur or small-business owner. One of the unfortunate trends in the late 20th century is corporate buyouts and accompanying downsizing. Recent mood swings in the U.S. economy in general and science and technology industries in particular suggest that mergers and buyouts will continue to be a business trend in the 21st century; and as a result, people will always be looking for new forms of employment. If these sorts of clouds have a silver lining, it is that some of those people will launch companies of their own, fitting into niches created by buyouts and the sloughing off of unproductive product lines and services. Those people (and you may be among them) will need to know something about marketing plans.

DEVELOPING A MARKETING PLAN

Developing the document we call a marketing or advertising plan begins with planning itself. Planning in this environment includes addressing the following issues:

- products and services
- prices
- distribution
- sales
- advertising
- promotion

Each of these topics is an important component of an overall marketing strategy. To begin with, you need to understand the market in which your product or service competes. What is its position in those markets compared with your competitors? What factors have led to the product's success? And what factors are limiting its growth?

Once you begin to understand these broad areas, you can focus on the more specific matters of what you and your company can do to improve the sales of a product. The choices are broad, but limited.

You could improve the quality of the product, pointing out that it does whatever it is designed to do better than its competitors. You could increase the advertising for the product. You could reduce the price customers pay for the product. You could expand your sales efforts into new geographic areas. You could improve the guarantees and warranties for the product. You could create more options in the product line to meet changing customer demands. You could explore new ways of getting the product to the customer; these are called distribution channels. You could start a new promotional strategy, offering incentives for buying your product. Coupons, free gifts, and the like are examples of this strategy. And more and more of these opportunities are moving online to the social media platforms we use daily. But you would not necessarily do all of these things at once; more is not better in marketing, despite the appearance that occasionally that is what companies are engaged in. Rather, just as was the case with public relations, marketing communications and the strategies inherent in developing them are highly focused.

Once you have made some decisions regarding the strategies you plan to undertake regarding your product, you will need to decide specifically how those strategies will be executed. Most important in this decision process is developing a program by which the success of your efforts can be measured. It is not enough to state that your marketing goal is to increase sales. Certainly you can tell if that is happening, but more precise statements of goals allow more precise evaluations of the success of your efforts. Instead of a blanket statement about increasing sales, a marketing goal should be stated more as follows: "To increase the sales of product X by *n* percent over a three-month period in geographic region Y." Now you can tell if the money your company is spending on advertising and marketing is a good investment. Did you succeed? Did you get close? If so, why? And if not, why not?

If, after considering all of these issues, you decide that you should develop a marketing and advertising plan, the following guidelines will help.

A marketing and advertising plan is built around three components: the market review, the marketing strategy, and the advertising strategy.

Market Review

Any successful marketing effort begins with a thorough understanding of the market a product is placed in. Consider the following questions, each of which could be the section of a report:

- **Growth.** Examine the sales performance for the past several years of the market your product or service will compete in. Such an analysis helps identify potential niches for your product or service.
- **Geographic particulars.** Examine the demographics of the region your product or service is to be sold in. Is the economy good? Do people have money to spend on what you are selling? Do they perceive a need for that product or service? Is your product targeted toward the predominant age, class, and education level within that region? Information on these matters will help determine the type of advertising that is most appropriate and where it should appear. For example, if you believe your product or service will appeal to people who commute to work by car, you probably want to buy some advertising spots on "drive-time" radio.
- **Sales performance.** Examine how your product performed last year vis-a-vis your predictions for its performance. Did it meet your expectations?
- **Market share.** How does your product or service stack up against its competitors? This question is usually answered in percentages.
- **Advertising spending.** Examine how much is being spent on advertising your product or service versus its competitors. (Check trade publications such as *Advertising Age.*)
- **Product tests.** Examine how well your product and its competitors have performed in controlled tests. Include the results of field tests, beta tests, customer service reports, and so on.
- **Marketing and advertising research.** If any formalized research was conducted on your product or its competitors, examine it and include the results.
- **Conclusions.** List the specific conclusions your market review has uncovered which will directly apply to future marketing and advertising. These should be organized as problems and accompanying opportunities. Remember, in the field of marketing, a problem is an opportunity.

Marketing Strategy

The second component of a marketing and advertising plan is the marketing strategy. It has two parts, based on the results of the market review:

- **Objectives.** This section is where you state specifically your sales projections for your product or service over a specific future time for a particular geographic region in terms of the percentage of improvement expected compared with the product or service's performance in that region for a similar time in the past. That is a long-winded way of saying something like: "Over the next three months, we expect the sales of consumer market drones in New England to increase by 15 percent."
- **Marketing strategy.** In this section of the report, you explain the activities—in detail—that you will undertake to make the objectives become a reality. These might include activities involving the development of the product, its pricing, its distribution, its sales, and/or its promotion. These issues lead directly to your advertising strategies.

Advertising Strategy

Once you have decided on specific marketing goals and objectives, you develop advertising strategies to enable your company to achieve those goals. This part of a marketing and advertising plan report has several sections:

- **Objectives.** Advertising objectives are similar to marketing objectives, but they tend to focus more on the emotive aspect of purchasing than on the hard statistics of analysis. For example, you might decide that you want customers to have an increased awareness of your product's quality. Or you might want them to trade up to a better model of what your company sells. You might want them to increase the frequency with which they use your product or service. Each of these topics can become the theme of an advertisement.
- **Advertising strategy.** In this section, you identify the specific activities you will undertake in advertising your product or service. This will include such issues as duration, or how long you will maintain a particular advertisement; and placement, or when and where you plan to run ads.
- **Creative strategy.** This section is where the art of advertising begins. Decisions made here are what contribute to the memorability of an advertisement, and that is the goal for anyone who advertises a product or service. For example, you might decide that the principal benefit of what you are selling is that it "is easy to use." Those four innocuous words

describe the user interface that launched Apple Computers. Once you have identified a principal benefit, you need to identify the reason that it is true. In the Apple Computer example, for instance, it would be that your customer can click on easily identifiable icons rather than remembering arcane codes, which was necessary for all computers before the first Mac hit the market. Principal benefits and reasons why they are benefits form the basis for creating an ad.

- **Media strategy.** Advertisements can be run in print media, on radio, on television, on billboards, on posters, on banners pulled behind airplanes, on the Internet, in social media, and in a variety of other places. Selecting when and where to run ads is building a media strategy. The media strategy should be closely aligned with what you learned about the demographics of your customer base when you conducted the market review. If your customers don't listen to National Public Radio (NPR), it makes no sense to waste any of your precious advertising budget placing ads (which NPR calls sponsorships) there, no matter how much you might like it.
- **Promotions.** In this section of the report, you will want to identify and analyze any sales promotions programs you plan to use. Be prepared to defend them.
- **Special activities.** Do you plan to participate in a trade show? Will you be testing a new product? These are special activities in a marketing and advertising sense. You need to identify your plans for these activities explaining why you are doing them and what you hope to accomplish with them.
- **Budget.** This section is self-evident. You need to develop a specific line-item budget for your advertising activities. Carefully weigh the costs of placing ads with their anticipated benefit regarding your marketing goals. Always allow a 5–10 percent contingency; Murphy's Law is a reality.
- **Calendar.** This is a schedule for what has become your advertising campaign. It includes when and where you place ads, as well as how long they run. It also includes any special events or promotional activities that will take place.

PRINT ADS

You might be thinking: "What does this have to do with me?" Absolutely nothing . . . if you work for a large organization that either outsources ad design or has graphic designers on staff. But what if you work for a small, entrepreneurial company with neither the budget to hire a freelance designer or the budget and space to keep someone full-time to do these activities? The answer is often: "Let's have the writer do it." So, if you find yourself in this situation, what follows is an overview of how to design print ads that work.

I think of print ads as a form of creativity that brings information to life. Even if you do not consider yourself to be a "visual person," there are strategies and standards you should be aware of in creating a simple print ad.

Print ads are made up of the following components: headline, body copy, and slogan (the copywriting part of an ad) and the visual design (ideally, the product of a graphic designer).

All ads begin with a concept. Think of this as the basic message that a print ad communicates to readers. Concepts should tie the headline to the visuals in an ad, so the concept is the controlling core of a print ad's design. Ad people often refer to this as "the big idea," but that is just gee-whiz talk for "thesis" or main point. All ads have them. A concept should stop the reader. What this means is that the concept is the central idea that is worked out through the headline and the visual component in such a way that readers cannot not look at it. Think of the ads that do this to you, even if you are casually flipping through a magazine trying to take your mind off the upcoming root canal at the dentist's office. When that happens, the concept worked. But it goes beyond this. Once a reader stops, he or she must be rewarded. This means that you want to write the body copy so that readers are drawn into the more detailed part of your message (but without using too many words—more about that a little later). The excitement, or reward, that a reader receives *must* come from the product or service that is being advertised.

Some ads are referred to as "copy stories." This means that the headline is bold and prominent, and that its wording leads readers right into the copy or text that explains the product or service in a compelling way. But other ads rely more on the visual component; usually this is done when the product is eye-catching, or there is some other visually appealing aspect to the ad. And like it or not, research proves that time and time again, attractive models—both male and female—stop readers at an ad.

There are several types of headlines:

- A direct headline states a benefit of the product or service (Example: "XYZ Generator Provides Hours of Backup Power in an Emergency")
- An indirect headline engages readers by pointing to a problem that the product or service will solve (Example: look at any erectile dysfunction ad)
- A question headline directly involves the reader (Example: "Feeling Sluggish?")
- A news-oriented headline suggests that the product or service is so important that it is newsworthy (Example: "Research Shows that ABC Lowers Cholesterol with No Side Effects")

Once readers are through with the headline, if you are successful, they will read the body copy. This should expand on the concept, often stated or implied in the headline. This is the time to both describe the product, its benefits, etc., and sell the product or the image of using it. Just like headlines, there are several types of body copy, depending on what you want to accomplish:

- Reason-why copy—just what the name says. This explains reasons why consumers should need or want the product or service.
- Dialogue copy—this is a little more challenging (and not as common). To use this approach, you must create characters for the ad and have them talk about the product or service.
- Narrative copy—more common than dialogue, but not as common as reason-why. With this, you have a narrator telling a short story about the product or service . . . in a compelling way.

After deciding on the type of copy, you need to decide how much of it is necessary. Some guidelines:

- Use fewer words (and more visual material) for image ads. These ads show attractive people using the product or service. Tobacco ads were notorious for this.
- Use fewer words for convenience products and services.
- Use fewer words for familiar products or services.
- Use more words to encourage readers to take action (buy the product).
- Use more words for products or services (if readers don't need it, you are going to have to spend time convincing them).
- Use more words for new or unfamiliar products.

The final text component of print ads is the slogan. This is one of the most challenging types of writing that exists in any discipline. If you have a talent for writing effective slogans, you can "write your ticket" and retire early. Already, some are probably coming to your mind ("Just Do It" or "The Pepsi Generation"). Good slogans are memorable, and the slogan is always placed in the lower right hand corner of an ad. It is the last thing you want a reader to see, and what you want the reader to remember and associate with your product or service.

The visual component of ads captured by how you design the ad layout, or overall appearance of the ad. Graphic designers rough out dozens of quick sketches, called thumbnails, that arrange different parts of the ad in different locations within the ad. Play with this. Try a lot of different layouts, but keep in mind the following four criteria:

- Balance—this is how the ad "feels" in terms of the weight of information left and right and up and down. Balance can be static, with approximately equal amounts of material in the four quadrants of an ad; or it can be dynamic, with more material in one quadrant than in the other three. Usually, dynamic balance weights the lower right quadrant of the ad.

- Unity—all visual components of an ad must complement each other so that the overall effect is one visual message.
- Movement—yes, information should "move" even in a print ad. You accomplish this by creating a dynamic balance that teases the readers eye through the entire ad, leading to the slogan in the bottom right corner. The following illustration, Figure 22.1, depicts the normal eye movement through a print ad; it is known as the Gutenberg Diagonal, which we also saw in Chapter 19. Any information—text or visual—that is placed on or near this line will be noticed more and remembered more than information placed away from it.
- Color—to explain this would require a lengthy discussion of color wheels and color theory. Suffice it to say, that if you combine colors and you or someone else thinks the result is awful, change the colors. This has been called a "felt sense"—in other words, all of us have a deeply felt sense of whether something is working visual. If you don't believe this, ask a friend or partner to hang a picture on a wall for you.

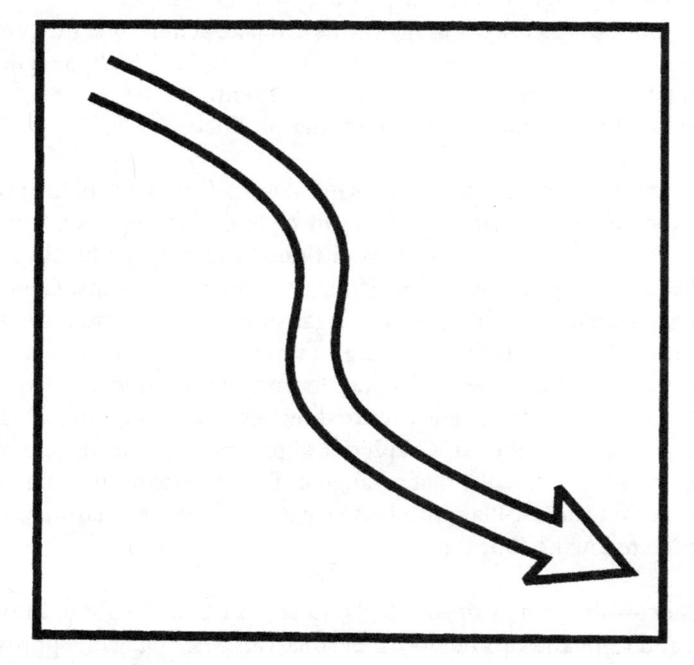

Figure 22.1. Gutenberg Diagonal.

The information in this section certainly will not enable you to be a professional ad designer, but if you experiment with it, you will develop the ability not to embarrass yourself if you ever need to create a print ad. Plus, if you do work for a large organization, this information will enable you and the graphic designer to "speak the same language" in coming up with ideas for effective ads.

WEB ADS

Like print ads, you might find yourself in a situation in which you need to design an ad for your organizations website, or for some other website that you have identified your customers might see. The challenge to designing effective web ads is that everyone hates them, installs software to block them, clicks on a little "x" to close them, etc. But if these forms of persuasive communication are a necessary evil, you should at least understand the strategies that (supposedly) make them work.

Web ads attempt to do any (and perhaps all) of the following:

- create awareness about a product or service
- generate interest in a product or service
- disseminate information to readers about a product or service
- create a positive image about the product or service
- create a strong brand sense for readers
- stimulate product trials

That last one is the most important one. The advantage to creating an ad for the web is that it is an interactive medium. If you are persuasive enough, readers choose to do something—order the product, etc.

The most common type of web ad is the banner ad. These are located around the edges of a web page—top, sides, bottom. The most effective location is the top of the web page; the least effect is the bottom. If you are going to design a web ad, re-read the print ad section. All the criteria for designing print ads apply to web ads. But for web ads, the strategies that involve movement mean actual movement, rather than creating a sense of movement through placement of information. Web ads can incorporate video (which if you are handy with a relatively decent camera you can capture) or animation (which is probably beyond your design capabilities, as it is mine).

What you want to remember about web ads is that your organization will be more likely to want these than print ads, so you might have more of an opportunity to design one. That being so, the image of your organization can be more dependent on web ads. So, apply the design strategies wisely.

CONCLUSION

In this chapter, like the one before it on public relations, we have explored a brief overview of an entire field—marketing and advertising. The information presented here serves several purposes. It provides a foundation for you to be able to talk with professional marketers or advertisers if you decide to contract these activities outside your company. It also provides a basis for developing some of these materials yourself, if you are an entrepreneur or a small-business owner. And finally, it provides an introductory understanding of key concepts that are treated in detail by the suggested readings listed next.

SUGGESTED READINGS

Parente, Donald, and Kirsten Strausbaugh-Hutchinson. *Advertising Campaign Strategy: A Guide to Marketing Communication Plans,* 5th ed. Mason, OH: South-Western College, 2014.

Shaw, Mark. *Copywriting: Successful Writing for Design, Advertising and Marketing.* London: Laurence King, 2012.

CHAPTER 23

How to Design Training Programs

Sooner or later, an executive at your organization is going to recognize that you are an excellent writer, as well as a more than competent speaker and presenter. As a result, you are going to be asked to develop training sessions for your company. In fact, you probably already do this, even if only informally. All of us teach people how to do things on an almost daily basis; but perhaps we have not had the opportunity to design and present formal training. So, in this chapter, you will find standard, time-tested strategies on how to develop formal training sessions. And if you never get the opportunity to do such a thing, realize that everything this chapter contains also applies to the informal training you do all the time.

INTRODUCTION TO TRAINING

Training, when done well, can be thought of as a method for encouraging professional development in organizations. It is a process for developing skills so that employees can more effectively perform specific jobs and tasks. As such, training focuses carefully on skills development, or the ability to do something, rather than focusing exclusively on knowledge development, or the nature of knowing something. Obviously, these are not mutually exclusive. In order to do something well, one must know a great deal about it. But the issue here is one of focus, and the focus of corporate training should be on activities that improve the productivity of an organization.

Training, therefore, develops skills for specific professional activities. It emphasizes doing, and appropriately designed, it provides opportunities for professionals to attain new levels of skill attainment. Think of training as a "closed system"—there are right ways and wrong ways of doing things within a professional environment. As a result, training emphasizes prescribed methods of performing one's job-related tasks. The intended result is for training to achieve positive organizational change due to the new repertory of skills and behaviors that were not present prior to the training and that the training provided.

Training is often contracted through consultants or consulting organizations. There are good reasons for this—objectivity of an outsider, who is unaffected by organizational politics, for example. But training can effectively be designed and offered in-house, as well.

Effective training, therefore, should be thought of as needs-centered training. Designing effective training requires the following:

- analyze the training task
- develop training objectives
- organize training content
- determine training methods
- select training resources
- deliver training
- assess training

Needs Assessment

Analyzing the training task requires performing a training needs assessment. Think of this as the evaluation of an organizational deficiency. Virtually all training is the result of someone noticing an aspect of an organization that needs improving; in other words, it is deficient at some level in its present state. The needs assessment analyzes that deficiency, what it affects, who it affects, and how it affects the organization's productivity.

Conducting a needs assessment is simply the process of identifying what groups of employees do not yet know or cannot yet do, in comparison with the skills and knowledge that they already possess. And it fits all this into what the organization needs for improved functioning going forward. Needs assessments can pinpoint problems that training can solve. It can confirm that these problems actually exist. It can develop solutions that allow employees and management to solve or manage problems. And it can ensure that the training enhances both individuals and the organization as a whole.

Needs assessments can be conducted through surveys, interviews, or observations. Often, one can simply design a brief questionnaire that is shared with potential employee trainees and management; the questionnaire is designed to have both groups identify areas of expertise that would be useful to improve. Interviews can then be used as a follow up with representatives from each group to check on whether or not questionnaire answers are being interpreted appropriately for the good of the organization. In some cases, observing employees at work is useful to determine their efficiency and effectiveness at certain tasks and responsibilities. Whatever methods, or combinations of methods, are used for needs assessments, it is vitally important that all training design begins with this step. Otherwise, training will be unfocussed, and trainees are likely to find it more of an imposition on what they are doing than a help.

Training Objectives

Training objectives describe the goals of the training being designed and developed. They are written statements that identify what the training *should* accomplish. I emphasized the word, "should," because it is the most important aspect of developing training objectives. As trainers, you cannot promise what the training "will" accomplish. It might not accomplish it—for reasons completely beyond your control (employee motivation, for example, or the willingness of management to commit the time and resources to enact and support the training results). So, all you can promise in writing training objectives is what the training should accomplish.

Objectives should meet four important criteria:

- They should be observable. In other words, after training it should be obvious to anyone examining post-training behaviors if trainees are successfully applying what they have been taught.
- They should be measurable. In other words, a training objective must be written in such a way that post-training behaviors can be quantified. For example, this might be in terms of improvement in speed, accuracy, etc., depending upon the type of training being performed.
- They should be attainable. In other words, training objectives must be within the reach of trainees. If they are not, the training will not be successful.
- They should be specific. In other words, avoid writing objectives that are overly general; focusing the objectives on specific performance-related activities should be the goal.

As long as you are designing training so that it meets the needs discovered in the training needs analysis, the objectives you create will work, if they meet these criteria.

Designing Curriculum

When you design training, you are in essence creating and teaching a course. Plan on teaching skills in chronological order; this simply is common sense, because it replicates the order in which we perform those skills. Also, teach simple skills before more complex skills. Doing this establishes a base line of understanding for trainees, and they can use that as a foundation for becoming more proficient and more knowledgeable, as a result of the training you provide.

At this point in the training design, you will also need to decide on which training methods are most appropriate to your content and your audience. Basically, your choices are to use lecture, experiential activities, or facilitated group discussions. In fact, you are most likely going to use all three.

Lectures are economical. The trainer can present a lot of material in a short time. Lectures also provide a great deal of trainer control of the content, and they can be surprisingly flexible, if the speaker is confident and experienced. I have been designing and delivering consultant training to organizations for over 35 years, and I regularly adjust lectures based on the questions trainees ask me. Despite these important advantages, there are some notable disadvantages to lectures, as well. They can be too trainer centered. The focus of effective training should be on the trainees, not on the ego of the person delivering the training. And of course, some people are simply boring. If you are not an engaging speaker, you should limit your use of lecture to the minimum that is necessary, and focus on the other two methods for delivering content.

Experiential activities include case studies that trainees can perform and discuss. They also include simulations, project-based learning, and demonstrations—either by the trainer or by the trainee. The advantages of experiential activities are important. Such activities tend to be very engaging for trainees, and the experience of applying training content directly is good for increasing trainees' confidence in applying the content later in their work environments. As a result, experiential activities are excellent for enabling trainees to transfer content to the workplace. But just as there were some disadvantages to lectures, there are also some disadvantages to experiential activities. If an activity is under-designed, it will not be effective; so, plan and create activities accordingly. But be sure that in aiming for engaging trainees you do not create activities that are gimmicky or artificial. And especially, be wary of creating activities that can be viewed as threatening to trainees. All trainees are in training because they have a deficiency of some sort related to their jobs. They know this—especially if they have been required by their superiors to participate in training. So, make sure that experiential activities improve trainees abilities without needlessly exposing their shortcomings.

Facilitated group discussions are an effective strategy for focusing content after it has been delivered through lectures and/or experiential activities. In using it, one should encourage all trainees to participate. The trainer should describe summaries of what trainees say, and probe responses. Be sure to monitor nonverbal behavior; often that will say a lot about how trainees are comprehending training content and thinking about how to apply it. And try to draw out minority opinions, disagreement, etc., so that these are not taken back to professional responsibilities unresolved.

No matter the training methods chosen, trainers also have the responsibility of selecting and managing resources that make these methods work. These resources can include handouts that summarize lectures, printed case studies, PowerPoint slides, or suggested additional reading and research. Virtually anything that makes training a more effective experience for trainees is a resource to be considered.

Assessment

Once training has been delivered and is complete, it is vitally important to assess its effectiveness. Only through assessment, analysis, reflection, and adjustment can training be an ongoing method of improving productivity for an organization. What you are trying to determine is if the training worked. If it did, what made it successful, so that it can be applied again in the future? And if it did not succeed, even partially, what can be avoided or redesigned for future applications in training?

There are numerous methods for assessing training, dependent upon the type of training that was offered. For example, one can observe the training content being applied on the job; certain types of training that are focused on specific activities lend themselves well to this.

But others do not. In those cases, it might be more useful to develop a questionnaire that encourages the trainee to assess what he or she has learned. The easiest way to accomplish this is to use a Likert scale. This is simply a series of statements that describe training content plus a 1–5 scale for responders to use in order for them to assess the effectiveness of the content described in each statement. Here is a common example:

The content of the training can be applied in my daily work.
1—strongly disagree 2—disagree 3—neither 4—agree 5—strongly agree

CONCLUSION

This chapter has presented an overview of training strategies that can assist in the design, development, and presentation of formal training in corporate environments. But even if any training you ever do is more informal than that described in this chapter, remember that all of the concepts apply. You still need to figure out what an audience needs from you if you are teaching them how to do something, even if it is simple. You have to determine what the goal of your instruction should be. You have to determine the best ways to teach them. And you have to figure out if they "got it." Everything in this chapter can help you do this.

SUGGESTED READINGS

ASTD Trainer's Sourcebook (varied subjects). New York: McGraw-Hill, 2000.

Meier, Dave. *The Accelerated Learning Handbook: A Creative Guide to Designing and Delivering Faster, More Effective Training Programs.* New York: McGraw-Hill, 2000.

PART VII

Finishing Your Work

CHAPTER 24

How to Avoid Common Writing Problems

Writing is one of the most difficult tasks anyone in any profession has to do. But this isn't news, is it? That is one of the reasons you are reading this book—to learn ways to make writing easier. What makes writing difficult is that it is creative, and creative processes take time. It is also hard work. Now, the notion that scientific or technical writing is creative may be foreign to you. But they are. You must create ideas to be communicated, words to carry those ideas, sentences and paragraphs that will be clear to readers, and overall organizational schemes that will make the logic of your ideas apparent. No wonder it's difficult.

Among the general difficulties of writing, however, are some common problems that occur regularly and that can be avoided. These are writer's block, organizational problems, punctuation problems, readability problems, and style problems. Every writer who has ever written—from the best-known novelist to you and me—has had to work through these problems. For some, using trial and error, it takes a long time, and it is frustrating work. For others, using proven methods, it still takes time, but not as much of it. And it is no longer hopelessly frustrating.

WRITER'S BLOCK

Writer's block is what we call the experience of getting stuck while writing. Everyone has experienced it, so you're not alone. The writer writes and all of a sudden can go no further. Minutes and more minutes pass, sometimes adding up to days, and nothing gets done. Deadlines approach, and with them, panic. Although the results are the same, writer's block has several sources: lack of information, lack of a well-defined purpose, lack of a thesis, poorly analyzed audience, and lack of confidence. Fortunately, each of these problems has a solution.

Lack of Information

The solution to a lack of information is simple: get more of it. But naturally, it's not quite as easy as that. First, one has to identify the problem. The surest clue is if you find yourself writing in circles, being repetitious, and not getting anywhere.

After you have decided that you do not have enough information or the right kind of information, you will have to discover the cause. Is it because you have not researched your subject thoroughly, or is it because you have kept poor records of your information? If it is the result of faulty research, then you will have to stop writing and do more information gathering. If you find yourself doing this often, then you might correctly guess that you have a problem in defining your communication purpose. More about this concept later. If your lack of information is the result of poor records, then you will have to redo some of the research to refresh your memory. The lesson of this dilemma is clear: don't trust your memory—at all, with anything. If it is important, or if it seems it may ever be important, *write it down, or photograph it with a smart phone, etc.*

Regardless of the cause of your lack of information, when you discover that as the source of your writer's block, you must stop writing. That's right; stop writing. Continuing to write at this point will accomplish nothing positive, and you will have to rewrite it anyway. After you have found enough information, then you can begin writing again, with a new, clearer sense of purpose. The important skill to develop is the ability to notice that a lack of information is the source of your writer's block. Remember that being repetitious is often the clue.

Lack of a Well-Defined Purpose

A poorly defined purpose for your communication will inevitably make writing more difficult. It may block it altogether. Sometimes, you will experience this problem as a lack of information; often you will notice it in paragraphs, sections, or entire reports that shift topic in the middle. For any case of poorly defined purpose, the solution is simple. Write the following sentence:

"The purpose of this document is __________."

And fill in the blank. Every technical document has a purpose, and all the information you include in the document should advance that purpose. If you want to make your work even easier, write the purpose statement sentence *before you begin to write* the document. Then you can develop a three-part purpose statement introduction around that sentence, as well as outline your report making sure that every word relates to your purpose.

Lack of a Thesis

"Thesis" is simply communication jargon for "main point," the single most important point of information you want your readers to know as a result of reading your document. And every technical document has one of these, too. If you don't know what yours is, you are doomed to write rambling communications that will (perhaps) eventually get to a point. Of course, no one can use these things, so you're better off deciding what your main point is before you write. Then you can use all your information to develop it. The way you identify your main point is exactly the same way as you identify your purpose. Write the following sentence:

> *"The main point of this document is__________."*

And fill in the blank. Remember, though, that your main point is not your goal; you identified that as your purpose. The main point is what you want readers to know.

Poorly Analyzed Audience

Although audience analysis was already discussed in Chapter 3, it bears repeating that a poor grasp of your audience can be a source of writer's block. You can recognize this problem best through experience. You find yourself staring at your computer wondering who is going to use your report and what on earth do they want from it. If that ever happens, recognize that these two questions are vital audience analysis questions. They should be answered *before you start writing.*

So, the solution to a poorly analyzed audience is also to stop writing and go do the audience analysis. Only after having done it will you be less likely to waste time doing nothing.

Lack of Confidence

A lack of confidence in your abilities as a writer is a common source of writer's block, and it is usually self-fulfilling. If you do not think you can communicate effectively, then you will not be able to do so. Practicing systematic approaches to writing problems is a solution to this dilemma. This method also tends to be self-fulfilling. The more you follow a writing process that works, the more positive feedback you receive about your reports and papers, and consequently, the more confidence you will develop in your communication abilities.

Writer's block, however, may also come from outside pressures. Recent research into writer's block suggests that much of it stems from unrealistic expectations about writing and unrealistic demands placed on writers. One of the most common unrealistic expectations about writing is that it can be done quickly. It can't, not even by intuitive-perceptive types who successfully complete writing projects at the last moment. As pointed out earlier, these types work long and hard turning ideas over in their minds, and evaluating and discarding hypotheses long before they began to produce a document. Writing is arduous work, no matter how you prefer to do it. Because of the differences in our preferences for how we order our work, writing projects should be kept as flexible as possible. Certainly deadlines are a reality, but periodic milestones may even contribute to writers' block for certain people.

One other thing: if you find that you are stuck, whatever you do, don't continue to write. You will accomplish nothing. This advice may sound like an excuse to procrastinate, and if abused, it is. But if you encounter writer's block, stop writing for the time being, and work to solve the cause of your block. You will waste less time over the course of writing a report or paper or designing online information, and eventually you will develop ways to avoid the common causes of writer's block.

Moreover, don't forget that this applies to the use of e-mail and other forms of social media in professional environments. Writing is writing, regardless of the medium used to communicate.

ORGANIZATIONAL PROBLEMS

Organizational problems are discovered during the editing phase, which is what makes editing vitally important to the production of scientific and technical documents. And this is a particular problem of using e-mail for professional communications: unorganized writers can create more harm than good by sending an e-mail no one understands. So, in any medium, you want to ensure that organizational problems are found and fixed. If they are not, readers will experience them as complete breakdowns in the communication. They will have to puzzle out how you got from one point of information to another. No reader likes to do this, and most will not.

The solution to organizational problems has already been discussed at length in an earlier chapter.

PUNCTUATION PROBLEMS

Accurate punctuation does not ensure accurate communication, but accurate communication is tremendously enhanced by accurate punctuation. Think of punctuation as the road signs along the readers' course. Punctuation keeps readers oriented to where they are, hints at where they are going, and reduces confusion along the way. Take the following punctuation test, without first looking at the corrected version that follows, to see how well you know punctuation. The test focuses on the most common punctuation problems found in technical writing. When you finish, compare your results to the corrected version and give yourself one point for every point of punctuation that agrees. If you score 40 or better, you should not worry about your punctuation abilities. If you score less than 40, the punctuation problems depicted in this test are discussed later in the chapter.

By the way, it bears repeating some advice mentioned earlier in this book. Don't turn over the responsibility for this vitally important aspect of writing in professional environments to the programmers at Microsoft. There are several examples of punctuation and grammar suggestions in Word that are simply wrong. Better that you spend a few minutes learning how to use standard punctuation. Your readers will thank you for it.

Punctuation Test

Punctuate the following sentences. Do not change anything else, such as the capitalization or wording. Two of the sentences require you to choose the correct word.

1. The report was bad and the presentation was worse.
2. Five reports are required a proposal a set of instructions an interim report a presentation and an analytical report.
3. The programmers John Jones Erin Davidson and Will Watson were well qualified for the job.
4. Because the proposal was late the company lost the contract.
5. If we think we can find a solution.
6. In that case let's submit a proposal.
7. Since he published "The Future of Artificial Intelligence" Professor Ward has become famous.
8. Professor Ward wrote The Future of Artificial Intelligence
9. Professor Ward wrote The Future of Artificial Intelligence he has become famous as a result.
10. Walters report Virtual Reality and Smart Games was well received.

11. Its well known that the groups findings contradict its assumptions.
12. If we look at the diagram Figure 4 we can see the problem.
13. None of the reports (was/were) correct in (its/their) assumption.
14. A phoneme an element of sound is one of the building blocks of spoken language.
15. The finding of the seven different reports on extraterrestrial intelligences (was/were) made public today.
16. The reason our proposal was rejected is simple it was too expensive.
17. The manager rejected our proposal for the following reasons it was too expensive our conclusions were vague and the time schedule was too long.
18. Who wrote The Future of Artificial Intelligence
19. The feasibility study was inconclusive however the project will be done anyway.
20. The reports findings indicate that because the procedure was incorrectly followed it is impossible to fund the full scale project.

Corrected Version of the Test

1. The report was bad, and the presentation was worse.
2. Five reports are required: a proposal, a set of instructions, an interim report, a presentation, and an analytical report.
3. The programmers—John Jones, Erin Davidson, and Will Watson—were well qualified for the job.
4. Because the proposal was late, the company lost the contract.
5. If we think, we can find a solution.
6. In that case let's submit a proposal.
7. Since he published "Virtual Reality and Smart Games," Professor Ward has become famous.
8. Professor Ward wrote "The Future of Artificial Intelligence."
9. Professor Ward wrote "The Future of Artificial Intelligence"; he has become famous as a result.
10. Walter's report, "Asynchronous I/O in Micro-computers," was well received.
11. It's well known that the group's findings contradict its assumptions.
12. If we look at the diagram (Figure 4), we can see the problem.
13. None of the reports was correct in its assumption.
14. A phoneme, an element of sound, is one of the building blocks of spoken language.
15. The finding of the seven different reports on extraterrestrial intelligences was made public today.

16. The reason our proposal was rejected is simple: it was too expensive.
17. The manager rejected our proposal for the following reasons: it was too expensive; our conclusions were vague; and the time schedule was too long.
18. Who wrote "The Future of Artificial Intelligence"?
19. The feasibility study was inconclusive; however, the project will be done anyway.
20. The report's findings indicate that, because the procedure was incorrectly followed, it is impossible to fund the full-scale project.

This test covers all the common punctuation problems (and some of the grammatical problems) writers are likely to face. What is not covered here can easily be found in a good handbook on punctuation. One of the misconceptions about punctuation, and language use in general, is that there are a thousand rules with ten thousand exceptions. This assumption is not exactly accurate. There are some rules, and there are some exceptions, but not enough of each to preclude being able to remember them. The rest of this section will explain what the test covered. After you have compared your answers with the correct ones, you can see how much or little you need to commit to memory.

Sentence 1 exhibits a common problem concerning punctuation: the use of commas. When two complete sentences are joined with any of these words (and, or, so, but, for, yet, nor), a comma is always required to separate the sentences.

Sentence 2 deals with the use of colons (:). Colons are used to set off lists from the rest of a sentence. Commas are used to separate the items of the list. The comma before "and" is optional. But leaving it out often leads to confusion and misreading. Put it in.

Sentence 3 displays ways to insert nonessential material into sentences. Usually, writers can do this in any one of three ways: with commas, with dashes, or with parentheses. In this case, however, the material that is inserted is in the form of a list. Because it contains commas, you should use either dashes or parentheses to avoid confusion.

Sentences 4, 5, and 6 deal with introductory material placed at the beginning of the sentences. If the introductory material itself contains a sentence after the transitional word ("Because" or "If"), a comma is required. Otherwise, as in the case of sentence 6, no comma is necessary unless you want the reader to experience a definite pause. Placing the comma in or leaving it out has nothing to do with the length of the introductory material.

The important thing about sentences 7, 8, 9, and 10 is the treatment of quotation marks in relation to other marks of punctuation. Periods and commas are *always* placed inside quotation marks. I know some of you are saying, "But that's not the way I was taught." It's not the way I was taught, either, but the English punctuation system is dynamic. It changes. And the changes are toward greater simplification. Semicolons are placed inside the quotation mark if they are part of the material being quoted (for instance, a title); otherwise, they are placed after the quotation mark. The same thing holds true for colons, dashes, question marks, and exclamation marks.

Sentence 11 exhibits the common problem of what to do with apostrophes. It seems most people ignore them, and this sentence is an example of what can come of doing that. The first word of this sentence is a contraction of "it is." The apostrophe is *always* included. The word "group's" is a possessive noun. The apostrophe is included here, too, to show that something belongs to the group instead of there being more than one group, a plural. The next-to-the-last word in the sentence is the possessive form of the pronoun "it." I realize this is confusing, but the possessive form of "it" *never* has an apostrophe. This rule is just something that all of us simply have to commit to memory.

Sentence 12 is a version of sentences 4 and 5. The difference is the inclusion of "Figure 4." If you place the parentheses around this term, you have to lift the comma and place it *after* the final parenthesis to show that the term goes with the introductory material and not with the rest of the sentence. Otherwise, the sentence will not make sense.

Sentence 13 depicts a source of confusion that has been in existence for two generations. It probably confused a number of you. The logically correct version of the test lists a form that sounds odd. The reason it does is that this is an example of the difference between conversational English and written English. Most people are no longer offended by "None were" in conversation. Increasingly, usage experts are saying that it is acceptable in written language, as well. Fifty years ago, it would have been thought an example of poor writing, however. "None was" is both correct and logical. "None" is the contracted version of "not one." This is an example of our language in flux. People who are more conservative in their approach to language usage continue to use "none was." Journalists are a good example of this. More and more people, however, are comfortable with "none were." And most of us will live to see it universally accepted over the next generation.

Sentence 14 is identical to sentence 3. Here, however, all the options are available to you. Be sure you choose among them for the right reasons, though. Communication research has suggested that dashes call more attention to the material between them than commas and that parentheses call less.

Sentence 15 is an example of how easy it is to be careless when writing. The subject of this sentence, "finding," is separated from the verb, "was," by eight words. Writers can easily forget whether the subject was singular or plural over that span. This type of confusion is one of the reasons that editing is important.

Sentence 16 exhibits a different use of the colons. Colons can be used to signal to the reader that information is coming up which will explain information that has just been read. This sentence and sentence 17 are examples of that.

Sentence 18 deals with coordinating quotation marks with other marks of punctuation. See the explanation of sentences 7, 8, 9, and 10.

Sentence 19 treats the punctuation of a different type of sentence-joining word. When sentences are joined with words such as "however," "therefore," "consequently," and about 40 other similar words, the word is preceded with a semicolon and followed by a comma. This punctuation signifies the longer or stronger pause these words require.

Sentence 20 also deals with punctuating material that is additive to the sentence. Here, however, the punctuation is entirely dependent upon whether the writer wants the reader to feel a pause. If not, it can be omitted entirely. One other aspect of this sentence is the hyphenation of "full-scale." Whenever two words are used together to modify another word, they become what is known as unit modifiers and require hyphenation.

This little exercise should let you know where you stand with regard to punctuation. If you did well, terrific. If not, solving the problem will not require as much work as you might have thought. See the suggested readings at the end of this chapter.

READABILITY PROBLEMS

Readability is a buzz word that most of us have heard. Much has been said about it, and numerous formulas have been developed over the past 75 years that supposedly test it. Readability is the likelihood that a projected audience will be able to read and comprehend a piece of documentation. It is an extremely important aspect of writing in any technical field, but particularly in the computer industry. One problem remains; however, no one has agreed yet on a readability formula that works, or if they are useful at all.

Let's look at two popular and widely used formulas, examining their strengths and weaknesses. If your word processor runs a readability analysis on your work, odds are that it uses one of these two formulas.

Gunning's Fog Index

This simple formula is aimed at locating the audience on a grade scale that is supposedly based on their formal education.

1. Select a segment of text that is approximately 100 words long, to the nearest period. For more accurate results, choose text from the middle of

a document. Introductions, leads, and conclusions usually exhibit slightly different styles that may skew the results of this test.

2. Count the number of sentences in the selected text.
3. Determine the average sentence length by dividing the number of words by the number of sentences.
4. Count the number of long words (those with 3 or more syllables). But don't count proper nouns, words that have 3 syllables because prefixes or suffixes have been added (for example, re-create-d), or words that are combinations of one- or two-syllable words (for example, storyboard).
5. Add the number of long words to the average sentence length.
6. Multiply this result by 0.4.
7. The result is the Fog Index. If you place it on a scale of 1–21, you will have the approximate number of years of formal education a reader would need to understand the document easily: 1–12 corresponds to grade and high school; 13–16 corresponds to college; 17–18 corresponds to a master's level college education; and 19–21 corresponds to a doctorate level.

NOTE: To increase the reliability of this formula, take several samplings from throughout a document.

Flesch Readability Scale

This test, like Gunning's Fog Index, is based on the length of words and sentences. It strives for greater accuracy by adding features that are analyzed, such as the total number of syllables. It also has the advantage of not being limited to a particular amount of text. Any size selection will do. In fact, several text analysis software packages include the Flesch Readability Scale as a provision to analyze entire documents. Here's how it works

1. Determine the following:
   a. total number of words (A)
   b. total number of sentences (B)
   c. total number of syllables (C)
2. Divide the total number of words (A) by the total number of sentences (B) to obtain the average sentence length (D).
   $A/B = D$
3. Multiply the result by 1.015.
   $D \times 1.015 = E$ (approximately 20)
4. Divide the number of syllables (C) by the number of words (A) to obtain the average word length.
   $C/A = F$

5. Multiply this result by 84.6.
   F x 84.6 = G (approximately 150)
6. Add E and G. E + G = H
7. Subtract H from 206.835.
   206.835 - H = Flesch Score
8. Place this score on the following scale.
   90–100—very easy
   80–90—easy
   70–80—fairly easy
   60–70—standard
   50–60—fairly difficult
   30–50—difficult
   0–30—very difficult

Although most explanations of the Flesch test also go into educational levels, such labels are not necessary. The scale does a fairly good job, as readability tests go, of explaining the relative difficulty of a document.

Both of these readability tests share a common fault. They oversimplify the task and the product of communication by ignoring the single most important issue—audience. Communication cannot take place in a vacuum, and for that reason, it is very difficult (some would say impossible) to measure its effectiveness quantitatively. In addition, these tests are based on word length and sentence length. This ignores the integrity of the subject matter. Try writing a few paragraphs about liberty, independence, declarations, revolutions, and constitutions. Your results will run afoul of each of these readability formulas.

While it is generally true that shorter words and shorter sentences are easier to read, it does not follow that they are always easier to understand. Sometimes, the short, choppy nature of such a style leads readers to skim a document, not paying close attention to what they are doing. Short sentences and simple words are best used in summaries, introductions, leads, and conclusions. A blanket indictment of longer sentences robs our language of its stylistic richness. It can even lead to stylistic contrivances. Writers are better off not having to worry about scoring below some arbitrary number on a less-than-effective scale. But writers must consider the level of expertise of their readers. This consideration is why audience analysis based on the knowledge shared between reader and writer is more important than the presumed reading level of that audience.

WRITING STYLE PROBLEMS

One's writing style is also an editing-phase problem. At the draft phase, writers should be concerned about communicating subject matter. Style is secondary. But at the editing, polishing phase, style becomes a way of enhancing the communication and of making it more accessible.

Style is the result of sentence structure and word choice. Although you might not think so, sentence structure is simple to understand and use. There are four basic types:

- Subject-verb (SV)
  The programmer (S) quit (V).
- Subject-verb-object (SVO)
  The programmer (S) hit (V) the flat screen display (O).
- Subject-linking verb-complement (SLVC)
  The programmer (S) felt (LV) sick (C).
- Subject-verb-indirect object-object (SVIO)
  The programmer (S) gave (V) the engineer (I) a headache (O).

By itself, this list represents a lot of variety that writers can use in crafting a varied and interesting style. When we add to this basic structure ways to modify words and sentences, the possibilities become almost inexhaustible.

Simply put, sentences may be modified in three ways:

- Left-branching, or before the main part of the sentence
  When we add to this basic structure ways to modify words and sentences, the possibilities (S) become (LV) almost inexhaustible (C).
- Right-branching, or after the main part of the sentence
  Writers (S) should vary (V) sentence structure (O), allowing readers the opportunity to pause, digest what they have read, and go on.
- Mid-branching, or amid the main parts of the sentence
  Writers (S), if they are wise and experienced, vary (V) sentence structure (O).

It is the variety of sentence structures and lengths that makes for a readable style, not the overuse of any one. Even though you can vary your writing style in other ways, these seven options can be used to create 46 different types of sentences. That's plenty for most of us.

CONCLUSION

This chapter has explained a variety of writing problems that are potentially confusing to people who write about technology. Rather than take a purely prescriptive approach by listing an endless series of dos and don'ts along with countless exceptions to every rule, I have attempted to draw upon every writer's common sense when it comes to dealing with these problems. It is surprising how often that intuition works when we trust it.

SUGGESTED READINGS

Arnold, George. *Media Writer's Handbook: A Guide to Common Writing and Editing Problems,* 5th ed. New York: McGraw-Hill, 2008.

Glenn, Cheryl, and Loretta Gray. *The Hodges Harbrace Handbook,* 18th ed. Boston: Wadsworth, 2012.

CHAPTER 25

How to Edit and Revise Your Work

Editing and revising are the final steps in writing a report or article about science or technology. These processes ensure successful communication. Almost everyone who writes does some sort of editing and revising, but many do not do so systematically with a specific goal in mind. For most, the idea of what constitutes a good report or article is only a gut reaction to their own work. Something seems good or something seems to need more work. While this process may succeed for some writers who have practiced it for a long time, it is difficult to develop the skills that are required to make accurate gut-reaction criticisms of one's own work. There is a better, easier way.

This chapter will present some ways that writers can assess the quality of their work and the work of others. It will also introduce the concept of an editing buddy system.

DIVISION OF EDITING

Good editing divides the task of improving a document into a limited number of areas:

- organizational logic
- mechanical development of topics
- writer's style
- quality of the manuscript

Each of these areas is vital to completing a successful, usable document.

Organizational Logic

Organizational logic is the single most important issue in writing a report or article about technology. It is the first thing a person needs to look for when editing a document. The ideas in a report or article should be interrelated—seamlessly. They should form a sequence of information that will appear to be predictable or inevitable to readers. In other words, a topic or an idea should anticipate topics or ideas that follow it. As I have said before, a document should have no surprises and no areas of confusion that the readers have to puzzle out for themselves. Edgar Allan Poe, referring to the writing of short stories, gave some advice to writers that we can very well borrow for the writing of reports and articles about science and technology. He said that authors should include nothing that does not advance the topic toward its inevitable end. Anything else is a tangent that readers will wander down with the writer and become lost together.

Mechanical Development of Topics

How the writer develops a topic goes hand in hand with organizational logic. If the ideas are arranged in a logical order, development of the topic becomes an easy task.

When editing an article, look at the lead first. Is it effective? Does it bring readers into the topic? Is it interesting, attention getting? It has to be for the rest of the article to work.

When editing a report, look first at the introduction. Does it have three parts—background, a statement of the specific topic, and a statement of what the report will do for readers? If it doesn't, rewrite it!

For both reports and articles, look next at the middle. Does the discussion maintain and develop the topic as the lead or introduction suggested? Is the overall progression of ideas predictable? Are there unnecessary surprises?

Examine the paragraphs individually. Each paragraph should begin with a topic sentence that gives readers an idea of what the topic of the paragraph will be and how it relates to the overall topic and/or purpose of the document. Then the paragraph should develop that topic and that topic only. Be on the alert for tangents in a paragraph, and remove them.

Look to see if there is transition between paragraphs, even between sentences. Transition is what links the ideas together. Transition can be accomplished in three ways. The writer might repeat an important word in two adjoining paragraphs or sentences, as I did with "transition" in the first two sentences of this paragraph. Structure can accomplish transition as it has in the way this paragraph and the three before it begin with the command voice.

Finally, transitional words and phrases establish links between ideas in an obvious way, as "Finally" does for this sentence and as "When editing a report" does for the third paragraph of this section.

Examine the ending of the report or article. Is it written with an obvious purpose in mind? Review the chapter on exits and conclusions just to make sure that the purposes of conclusions are second nature to you. Will the ending leave readers with a sense of fullness and completion, a sense that you have satisfied their needs? It should.

At last, check the punctuation throughout the document. It should reflect the logical organization and interrelationship of ideas throughout the document. It should also be accurate and correct to presently accepted standards. If you do not know what those standards are, buying a good handbook on grammar and punctuation would be a wise investment, as are regular visits to reputable websites on the topic.

Writer's Style

Style is the writer's voice coming through the words and sentences. Style is inevitably structure, how writers arrange words and sentences in patterns that sound natural to them. Written style, however, differs slightly from conversational style. It is a bit more formal, a bit more planned and measured. When we converse with someone, as opposed to when we make a planned presentation, we have little time to plan the arrangement of words and sentences so that what we say is clear. In conversation, we establish clarity by observing visual and oral cues from the other person to check whether we are being understood. But in writing, we do not have that luxury. We must anticipate readers' reactions, and we must be sure that the structure of our sentences and paragraphs represents the thoughts that are carried by them.

For example, are the sentences in a document varied in length and structure? If sentences are of nearly the same length and the same structure, the style will be monotonous and boring, regardless of the readers' interest in the topic. If necessary, review the chapter on solving common writing problems to see how sentences can be varied.

Check your choice of words. Will the words mean for your readers what you intend them to mean? Are they words the readers are familiar with? Whatever you do, don't choose words to impress your readers. You won't. So, if you regularly write with the thesaurus by your side, move away from it. Impress readers with the crystalline brilliance of your ideas, not with the pomposity of your vocabulary. If readers can't understand what you are trying to say, most will simply put down your document and go about what they see rightfully as more important work.

Is the style appropriate for the topic and for the audience? In this book, I have attempted a direct, informal style to create a sense that I might be talking to you in one of the professional seminars I hold in industry. You, the audience, and the topic determined that. And you probably did not believe me when I said that readers were the most important aspect of communication. Readers have power. They can refuse to read! If this book had been written for professors of communication, or as a textbook for a technical or science writing class, the way it "sounds" would have been inevitably different, perhaps more formal.

Finally, does the style convey the effect the writer intended? If you wish to sell an idea or product, then the style must be persuasive. If you wish to convey information only, such as in a report to superiors, then the style should appear to be objective, even though real objectivity is a fiction.

QUALITY OF THE MANUSCRIPT

The last thing a writer needs to examine when editing a document is the quality of the manuscript. This step is the icing on the cake, so to speak. It is the writer's last chance to polish the work.

A manuscript should be visually attractive. This means that it should be orderly, and that it should have obvious divisions and subdivisions. One of the best ways to test your manuscript for this quality is to lay it out on the floor and stand over it. Can you see clearly where the sections are divided? Are the headings surrounded by sufficient white space? If you cannot tell, revision is called for.

A manuscript should invite readership. Have you ever seen a report or paper that had the print crowded out to the edges of the paper from top to bottom and from side to side? Such a manuscript looks hard to read. The type should be clear in a manuscript, and the production should not be sloppy. Avoid dot matrix printers.

A report or paper should be sufficiently detailed. How much detail is sufficient detail? Only the writer can answer that accurately. There should be enough detail, though, to satisfy the projected readers' needs for information. Again, the answer is tied to audience analysis. Even though detail is vital, the document must be brief. These are not mutually exclusive criteria. If the writer has paid attention to logical organization and development of the topic, the document will be as detailed and as brief as it should be.

After you have done all this, proofread the document—preferably in hard copy rather than on a screen. Don't confuse proofreading with editing and revising, however. These are much more involved processes, and proofreading is only one small step. Editing and revising require being judgmental, asking whether a certain statement or organization or figure is as good as it can be, and then making it so. Proofreading is a quick check for simple errors. The best way to keep them straight is to remember that editing includes proofreading but that proofreading alone does not ensure editing and revising.

THE EDITING BUDDY SYSTEM

So far in this chapter, we have examined what writers should do when editing and revising documents. Now let's look at how they should do it.

The first thing you will want to do is to choose another person, your "buddy," whose critical judgment and honesty you can trust. You will agree with that person to edit each other's documents. One way to do this is to read your partner's document out loud to that partner. Every place you stumble, or hesitate unnaturally, signifies a place where the meaning is not as clear as it should be. Your partner makes note of this stumble and rewrites the troubled area later. When you have a document that needs editing, your partner returns the favor. This is an extraverted procedure. If the physical arrangement of your workplace makes it impossible to edit a document by reading it aloud or if your personality types make this sort of interaction uncomfortable, you can accomplish the same results with the following procedure.

Write out on a separate sheet of paper (or in an e-mail) a phrase or sentence that accurately describes the central point of each paragraph in your document. Give the document to your partner and have that person read it, doing the same thing for each paragraph. Compare the two lists. There will be differences. Discuss the differences with your partner. By doing so, you will gain insights as to how sections should be rewritten. This is an introverted procedure.

Finally, make sure that each document receives two types of edits: a technical edit, by an expert in the subject area, to ensure that the information is correct; and a usability edit, by someone who represents the audience's perspective, to ensure that the information can be understood.

CONCLUSION

In this chapter, we have looked at how writers can systematically edit and revise documents to ensure high quality. Editing and revising are vital, and the successful writer never overlooks them. In the high-tech industries, it seems that everything should have been done last week, or worse, last month. Nonetheless, take the extra day to edit and revise. The ends support the means.

To conclude this chapter, I've summarized the points of editing and revising in a checklist. Use it until it's memorized.

EDITING CHECKLIST

Organizational Logic

1. Are ideas related?
2. Is the sequence of ideas clear?
3. Does it make sense?

Mechanical Development of Topics

1. Does the lead or introduction work?
2. Is the lead interesting?
3. Does the lead attract attention?
4. Does the lead include background, a statement of the topic, and a statement of what the document will do for readers?
5. Does the discussion or body develop the topic according to what was said in the lead?
6. Does each paragraph begin with a topic sentence?
7. Does the paragraph develop the stated topic?
8. Is there transition between paragraphs?
9. Is there transition between sentences?
10. Are the words accurately chosen?
11. Does the ending work?
12. Does the ending tie down the subject?
13. Is the punctuation accurate?

Writer's Style

1. Is the style appropriate for the topic and audience?
2. Is the style varied?

Quality of the Manuscript

1. Is the manuscript orderly?
2. Are there enough headings and subheadings?

3. Are the headings clear?
4. Is the print easy to read?

SUGGESTED READINGS

Alred, Gerald, and Charles T. Brusaw. *The Handbook of Technical Writing*, 11th ed. Boston: Bedford/St. Martin's, 2015.

Amare, Nicole, and Barry Nowlin. *Technical Editing in the 21st Century*. Upper Saddle River, NJ: Pearson, 2010.

Rude, Carolyn, and Angela Eaton. *Technical Editing*, 5th ed. Boston: Allyn and Bacon, 2010.

PART VIII

Presentations and Meetings

CHAPTER 26

How to Make Professional Presentations

Many professionals in scientific and technological industries find that a significant number of papers and reports turn into presentations, but presentations are different from written reports in some important ways. When people read reports, they can pay attention or not, because they can always reread. If some part of a report is unclear, the readers can go back over it more slowly, attempting to puzzle it out. These are not excuses for sloppy writing, but they are realistic advantages of written communication. When an audience is listening to a presentation, they do not have these advantages. They hear the message only once, provided that the speaker is not terribly repetitious. They must focus on what is happening in the present, for if they ponder on what has just occurred, they are missing the present. They must pay close attention always because they cannot go back. They cannot skim or look ahead either. And if they have questions, often they are unwilling to ask them (although it is easier to ask a speaker a question than to ask a book). If you have ever sat through a poor presentation, you have no doubt experienced some or all of these difficulties.

Because of these limitations, presentations must be flawlessly clear. Rigid organization is a must. In order to understand the message, the audience must see how the speaker got from point A to point B for all the points in the presentation. This task is a big one, but fortunately I have a systematic way to approach it.

The system presented in this chapter will not make your presentations spellbinding—the content and your personality have something to do with that. But it will ensure that your presentations are focused on an intended audience, that the information is clear and logically developed, and that you do nothing to distract from the message. In other words, this process for making presentations

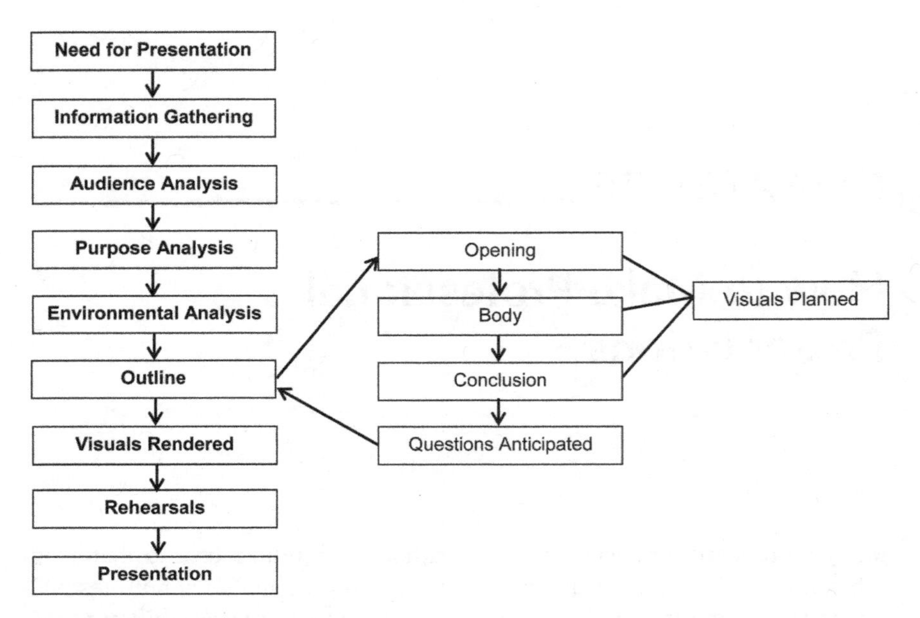

Figure 26.1. VOS for Presentations.

guarantees some level of success, but the polish and resulting accolades are up to you—the rewards of practice. Figure 26.1 depicts the process.

PURPOSE OF THE PRESENTATION

The first thing you must decide is the general purpose of your presentation. You have only two choices—to inform or to persuade. Of course, most persuasive technical presentations are also informative, or at least they should be. However, deciding the general purpose of your presentation is by itself not much help. The important step is the next one. You need to decide on a specific purpose for the presentation. In other words, are you going to inform the audience about topic x? Or are you going to persuade them to buy product y or to do project z?

Once you have decided, consider whether the presentation might also afford you a personal purpose—such as advancement, fame, money, and so on. Does any personal purpose that might apply to the presentation complement or contradict your specific purpose? As an example, let's imagine that you are scheduled to make a formal presentation before the Gaming Istanbul Digital Entertainment and Game Expo (GIST). During this presentation, you will announce the completion and successful beta testing of a new product line from your company. Such a presentation may serve to associate you with the new product line. This association is undoubtedly accurate, otherwise you would not have been selected to make the presentation. Such an association might affect your career path in a positive way—either by advancement within your company or by the offer of a new position within another company. Anything of this nature would be personal purposes that complement the specific purpose of your presentation.

Let's imagine another situation. You have been chosen to explain to a group of large, consistent, corporate customers why a promised product is already months behind schedule and apparently not as good as your chief competitor's newly released product, an activity known widely as "walking the plank." You realize that your company, and by association you, will be seen negatively. If your presentation doesn't work, it could cost your company millions of dollars in potential business; it could also cost you your reputation. This situation demonstrates personal purposes that are in contradiction to the specific purpose of your presentation—to inform an audience about a problem and convince them that it is not as bad as it seems.

As these examples show, the determination of purpose is extremely important when organizing a formal presentation. It determines the information you will select, the points you will make and not make. Purpose determines what you will say.

THE PRESENTATION AUDIENCE

Audience issues in presentations differ from those concerning written reports. Just as in written reports, it is important to know what level of expertise the audience has attained in your subject area, as well as whom they work for and what they do. But you should also consider the size of the audience. This consideration tells you something about the size of the room in which the presentation will be held, which affects the type of visuals you can use. (See Chapter 26.)

You should also consider these questions: What motivated the audience to attend your presentation? The subject matter? You? Were they required to attend? What specific objectives do they hope to realize by listening to you? The answers to these questions suggest ways to organize your information; they also suggest what should be included and what should be left out. For example, if your audience was coerced into attendance, you need to offer them something important or interesting early in your talk. If they chose to attend, seeking particular information, you can lead them to it a little more easily.

What does the audience know about you? Will they view you as credible or not? What information can you include that will enhance your credibility? These are extremely important questions if you are organizing a presentation before a large group at an industry conference. Realize that in such presentations you have about three minutes to convince the audience that you have something important to say before their attention drifts off, or in some instances, before they physically leave. For this reason, the opening of a presentation is very important.

What does the audience know about the specific topic of your presentation? Do they hold any emotional ties (pro or con) to the subject? For example, speaking about nuclear power is almost guaranteed to polarize audiences into camps. Knowing or suspecting the attitudes of an audience is therefore important. You can anticipate problems and defuse them before they become serious. This skill carries over very well into the running of meetings, where problems tend to be rife and attitudes strongly defended.

THE PRESENTATION ENVIRONMENT

The location of the presentation is another issue to consider. If possible, examine the space. Is it a large auditorium or a small conference room? Will the audience be seated theater-style or around tables? These considerations also affect the types of visuals you may use and your freedom of movement. They may even affect the formality of your presentation.

Does the presentation environment have potentially distracting elements such as lawnmowers running outside the window or humming fluorescent lights? Distracting elements can make it difficult for your audience to pay attention. Many distractions can be changed or avoided. If you encounter those that cannot, you will at least be prepared for them if you examine the environment prior to your presentation.

What arrangements will be made for you, the speaker? Will you have a podium, a microphone, a table, audio and visual equipment? These questions should be answered before you arrive to make a presentation. Remember, you will be nervous anyway; don't let unnecessary surprises make it worse.

Does your presentation have a time limit? The amount and type of preparation that goes into a 10-minute speech is vastly different than that required for a two-hour training presentation. And the 10-minute speech is more difficult! Will questions be allowed during the presentation or afterward? What will the audience be doing before and after the presentation?

These environment questions are certainly not as important as topic-related questions or purpose-related questions. But the environment for a presentation does affect its acceptance. Don't forget to consider it.

THE PRESENTATION TOPIC

There are some important areas to define concerning the topic of a presentation. For example, what do you know about the topic? And closely related to that: What do you need to know to make an effective presentation for your chosen audience?

Once you have gathered all the information you need to meet your specific purpose, you will have to organize it. Determine a limited number of main points to be made during the presentation, usually no more than five. These will be either points of information you want the audience to understand during an informative presentation or items of support for a persuasive presentation. Make sure that each point is clearly separate, that each point is similar in scope or level of importance, that they cover your topic, and that they do not overlap. For example, if you were planning a descriptive presentation about my MacBook Pro laptop, your main points might be (1) the CPU, (2) the keyboard, and (3) the display unit. You would not include the electric plug as a fourth point unless you wanted to make a humorous point about the importance of the power supply.

Once you have determined the main points to make, you will have to arrange them. Just as in written communication, this arrangement has to make some sort of sense to the audience. Its primary purpose, therefore, is to help the audience understand.

You might choose a chronological arrangement if your topic has something important to do with time, a procedural topic for example. You might choose a spatial arrangement if you are describing something physically. Usually this speech would be augmented with visuals.

Your topic might suggest an associational order of main points, such as cause and effect or problem-solution. Or it might require a multi-part order such as induction, deduction, familiar to unfamiliar, or simple to complex.

If the presentation is persuasive, you might opt for what is known as Monroe's Motivated Sequence. The first part of such a presentation attracts the attention of the audience and then shows the audience that they have a need, or a problem, that requires a solution. The presentation then moves on to supply the solution. Next, it shows how it would be if the solution were implemented. Then it finishes by outlining how to implement the solution. This type of presentation is a version of selling, convincing the audience members that they need something and then showing them how to get it. It is a powerful organizational strategy when the presentation purpose calls for it and when it is done right.

Although this section deals primarily with main points in a presentation, if the topic is broad enough and if you have enough time for your presentation, you can subdivide the main points. Only be certain that all the subpoints adhere to the criteria used in selecting and organizing the main points.

OUTLINING THE PRESENTATION

When outlining a presentation, it is important to follow a few widely used guidelines. First, make sure that each level of the outline is similar in scope. This rule applies to subpoints as well as to main points. Limit each section of the outline to one idea, using a short phrase to remind you of the point you want to make. This procedure makes the outline easier to follow if you use it as the notes for your presentation. Make sure that each section does not overlap, so that the audience is not confused. And if you subdivide a main point, the division should create at least two subpoints. Otherwise, it is like taking an orange, cutting it in half, and having a whole orange left—only smaller than the original.

Focus on transitions in the outline. They are what enable the audience to follow your train of thought. Realize that introductions are transitions. They connect the audience's not knowing about your specific topic to a readiness to be told the specifics of it. Transitions also should tie main points and subpoints together. Don't overlook conclusions. They are what brings a presentation to a successful completion. Each of us has been to a presentation in which the speaker stopped and the audience did not know whether the presentation was finished or not. Such presentations do not conclude; they just end. And even in the informal presentations made within your organization, avoid "that's-about-it" conclusions; those are not conclusions, they're surrenders.

TYPES OF DELIVERY

You can deliver a presentation in four possible ways, each with its advantages and disadvantages. Choose a method that makes you comfortable, but make certain it produces an effective presentation.

Manuscript Method

The manuscript method is what is often referred to as "reading a paper." It is most frequently seen in two situations: large, formal conferences that place rigid restrictions on the presentation's length and presentations in which the speaker is terrified to look at the audience.

One advantage to reading a manuscript is that all the speaker's words are determined beforehand. This reduces or eliminates the concern of getting lost or forgetting what one plans to say. It is advantageous for rigid time limits because the entire presentation can be practiced exactly as it will be delivered and timed to fit the requirement.

This method has disadvantages, as well. Most of us do not read out loud well without training and lots of practice. We tend to read in a monotone, with unnatural rhythms and hesitations. Reading a manuscript also precludes much eye contact with the audience, allowing the audience's attention to wander. If you are not paying attention to them, why should they pay attention to you? To some degree, you can get around this problem by writing notes in the margin to yourself saying, "LOOK UP!!!" Then, when you look up, place your thumb on the line you were reading so you can find your way back without tipping off the audience that you are lost.

Generally speaking, limit your use of the manuscript method to formal presentations that are required to be short. Remember that an eight-page, double-spaced manuscript takes almost 15 minutes to read; much more reading than that wears thin on the audience. Whatever your presentation is, do not resort to the manuscript method as a crutch. The disadvantages of the method are obvious enough to an audience without showing them your fear also.

Memory Method

If you were so inclined, you could memorize a presentation. The advantages are the same as they were for the manuscript method. In addition, memorization is also good for eye contact with the audience.

But the disadvantages are serious. First, it takes a long time to memorize eight pages of text for a typical 15-minute presentation. And if our reading aloud is bad, our recitation is worse—at least for those of us who lack formal training in this sort of thing. The reason for this is that we memorize not by committing a string of words to memory but by learning the rhythm patterns inherent in speech and hanging the words on those patterns. That is why a third-grade school play is delivered in the inevitable sing-song; it's how the lines were memorized: "ta-da, ta-da, ta-da, ta-da." To an extent, we adults are guilty of the same thing when we memorize a presentation. And the more we practice it, the worse it gets.

That's not the most serious disadvantage, however. The real danger is memory loss—forgetting what you were going to say, who you are, or why you're there. If that happens in a memorized presentation, you might as well sit down.

For these reasons, I cannot think of a single good excuse for memorizing a presentation.

Impromptu Method

This is the easiest way to prepare a presentation: don't prepare at all. Wing it; make it up as you go along; fake it. This method has no advantages, at least in terms of advancing your career. If you want to be fired from a company, however, the impromptu method of delivering a presentation just might do it for you.

But there is one unavoidable reason for making an impromptu presentation. You find yourself seated in the audience at a conference; the speaker to whom you are listening notices you are there and remembers you have done work in the area being discussed. Since the speaker has run out of intelligent things to say, he or she calls on you to stand up and share your knowledge with the group. At this point, trapped, you slowly get up, blush, stammer out a few unintelligent remarks, proving you know absolutely nothing about the topic or that you weren't paying attention, and sit down, vowing to get even with the speaker if it takes you the rest of your life.

The best thing that can be said about the impromptu method is, "Don't use it."

Extemporaneous Method

This method is preferred by most speakers. It means making a presentation from an outline or notes. It has all the advantages of good eye contact, naturalness of language, rhythm, pace, and voice modulation. It does require practice because it is easy to get caught up in conversing with an audience too long or to wander off the subject. These disadvantages notwithstanding, you should get in the habit of using the extemporaneous method most, if not all, of the time.

PRACTICING THE PRESENTATION

Practice is essential for success. Your objective should be a conversational delivery, and the practice should simulate the actual presentation. This means you should:

- practice out loud
- practice the entire presentation
- practice in the actual presentation environment, if possible, or in one that is similar
- practice using visuals
- practice movement and voice modulation

A participant in a seminar several years ago gave a presentation that demonstrated the importance of practice, especially the right kind of practice. In the interest of time and fairness to all participants, I had imposed a 10-minute limit for presentations. This speaker practiced long and diligently. But during his practice, when he reached a point he was particularly comfortable with, he would say "and so on and so on." When he made the talk, he had to fill in the "and so on's." When I stopped him at 10 minutes, he was a little over half finished with what would have been almost a 20-minute presentation.

VOCAL AND BODY DIMENSIONS OF A PRESENTATION

Many of the aspects of delivering a presentation are beyond the scope of this book, but are nonetheless important. This section will summarize a few of them, but you should consult the books at the end of the chapter for additional information.

Stress of the speaker's voice is important, because it can change the meaning of what is being said, as well as call attention to particular points in the presentation. Consider the following five sentences as an example. In each sentence, the stressed word is italicized.

1. The cat is in the washing machine.
   (No stress; simple statement of an observation)
2. The *cat* is in the washing machine.
   (As opposed to the gerbil)
3. The cat *is* in the washing machine.
   (As if you did not believe me the first time)
4. The cat is *in* the washing machine.
   (As opposed to being on top of it)
5. The cat is in the *washing machine.*
   (As opposed to being in the dryer)

In each sentence, the meaning changes because of the stress alone.

Volume is also important. The speaker must be capable of being heard by everyone in the room. This is one more reason why examining the presentation environment before making a presentation is important. If you have a light voice or poor projection, consider employing an amplification system (and practice using it).

Enunciation is important. The audience must be able to differentiate among your words. If you mumble or run words together, they will not be able to. This is another reason why the manuscript method should be avoided if possible. When reading a manuscript, you spend much of your time speaking to the podium.

You should also pay attention to the pace of your presentation. Realize that if you are anxious at all, you will automatically speak more quickly than normal. If things get completely out of control, your audience will hear a breathless, rushed presentation that is distracting solely by how it is delivered. Try to intersperse silent pauses in your delivery—not with "uhs," "ahs," and, worse, "ya-know?s"—but with momentary catch-breath silences in which you can collect your thoughts and the audience can digest information before you both proceed. At first when you try this, you will inevitably feel self-conscious, but with practice you can cure the filler habit. If you want a model for how this pausing is done well, observe a national network anchor reading the evening news. The silent pauses are natural, unobtrusive, but obviously there.

AVOIDING STAGE FRIGHT

Some degree of anxiety is a reality to all speakers. In fact, it may even be good for you; it gets adrenaline flowing and makes you seem excited about your message. But the important issue is to control it, as well as to remember that audiences are usually not aware of a speaker's anxiety.

If stage fright paralyzes you with fear, you can try to develop relaxation techniques. When you stand up to make a presentation, force yourself to look at the audience—not as a crowd but as individuals. Take a few slow breaths. This suggestion does not mean breathe so deeply you hyperventilate, but try to control your breathing and pulse as you start a presentation. Your audience will see what you are doing as getting organized before taking them into your material.

Adequate preparation is also an effective cure for stage fright. The more presentations you deliver, the more confidence you will have doing them.

CONCLUSION

In this chapter, we have examined methods for planning and delivering professional presentations. Practice is essential for success, as is a consistent focus on communication rather than performance.

But why is making presentations important? One reason is that they are a form of ritual; they provide beginning professionals with opportunities to demonstrate that they "belong to the club." In many ways, presentations are an important part of the corporate culture, enabling speakers to showcase their abilities and to reap the rewards of career advancement. Recent research suggests that speakers enter into implicit contracts with their profession and with the audience. In other words, by accepting an opportunity to speak to a group of our colleagues, we are agreeing to make detailed preparations that will support and demonstrate the value of our information, as well as agreeing to present that information in the best way possible. Many of the skills presented in this chapter lead to types of presentations covered in Chapter 27 "How to Run Effective Meetings."

SUGGESTED READINGS

Clapp, Christine. *Presenting at Work: A Guide to Public Speaking in Professional Contexts.* Washington, DC: Spoken with Authority, 2014.

Maxey, Cyndi. *Present Like Pro: The Field Guide to Mastering the Art of Business, Professional and Public Speaking.* New York: St. Martin's, 2006.

Maxey, Cyndi. *Speak Up: A Woman's Guide to Presenting Like a Pro.* New York: St. Martin's, 2008.

CHAPTER 27

How to Use Visuals with Presentations

Visuals are an important addition to presentations, which is why they were originally called visual aids. But that is all they are; they cannot and should not take the place of language in lengthy communications. Rather, they should be used to highlight presentations, to focus the audience's attention on important material. Correctly designed and used, visuals should support and expand the content of your presentation. They should clarify what you mean as you speak. If they are to do this, they must meet certain criteria:

- They must be visible, large enough for the whole audience to see—even those people who insist on sitting in the back row.
- They must be clear; their meaning must be obvious at a glance without explanation.
- They must be simple and easy for the audience to comprehend.
- They must be controllable, easy for you to use with your presentation.

By now, you are no doubt thinking that you have never seen visuals that met all these criteria, particularly the one about their meaning being obvious at a glance without explanation. How many speakers have we listened to who spent much of their time explaining their slides rather than giving us the information we came to receive? However, the fact that most presentations fall short of these standards is not a justification for ours to fall short of excellence as well. Good visuals polish a presentation. For that reason alone, good visuals should be our goal.

Many types of visuals are available for the speaker to use, each with advantages and disadvantages. In this chapter, the different types of visuals along with presentation methods will be examined with special attention paid to the type of presentation environment in which they are best used.

WORDS AND PHRASES

Words and phrases are points upon which the speaker will elaborate. They should represent important matters in the presentation, and they should be short and simple. Frequently, speakers will present the audience with an outline of their presentation on a PowerPoint slide at the beginning of the talk. Although this treatment is a little heavy-handed, it does work in that the audience is automatically oriented to the topic and to the order in which it will be treated.

Words and phrases might be presented through a range of media with equal success, as long as the criteria for the use of each of these media are also met. Just make sure that you do not force listeners to do too much reading; if they do, they won't be listening to you.

CARTOONS

Cartoons and animations are effective visuals when the speaker is dealing with a sensitive subject. This situation rarely occurs in the high-tech industries, but cartoons can also be used to depict people-oriented action in such a way as to enliven topics. These two uses are often useful when making a presentation about science and technology audiences that do not share the speaker's knowledge of the subject.

WHITE BOARDS

Despite all the recent advancements in presentation technology, white boards are still popular media for informal, small-group presentations. Regardless of their popularity, most people misuse them. If you are planning to depict information on a white board during a presentation, write as much of the information on the board before the presentation as you can. Then you can reference the information at the proper time during your presentation. Make sure that what you write on the board is simple and neat. Successful use of these boards requires a little practice; it isn't as easy as it looks. The first time you do it, you will notice how odd writing at a vertical angle is and how difficult it is to write legibly.

Once you are ready to reference the information to the audience, prime them. Let them know what is coming up and why it is important. Remember also that white boards should be limited to small presentation environments. People who are seated more than 30–40 feet from the board will have a hard time reading what you have written. And if you write too large, you will use up so much of the board that it will not be an effective visual medium. Remember: Make sure you have a dry marker to use instead of a magic marker. If you don't, you will create more problems for yourself than you can imagine—you can't erase magic marker from a white board.

Finally, whatever you do, avoid turning your back on the audience, writing on a board, and talking to what you have written. This is a guaranteed way to distract an audience.

POWERPOINT AND PREZI

PowerPoint is excellent for presenting information to large audiences. It is easy to organize the entire presentation visually, and use it repeatedly. Remote control is another advantage in that it allows you the freedom to move about in your presentation, making the event more dynamic than static

Effective use of PowerPoint requires a different set of criteria. First, if you are presenting in a completely darkened room, be sure that you light yourself at the podium, and perhaps resist the temptation to wander about the presentation space in the dark. Some people find it difficult to pay attention to a disembodied voice in the darkness; they are likely to take a nap. Second, prime the audience. Let them know what each slide means and what each segment of slides means, especially if they combine visuals and text. Third, tell them why each slide is important.

Prezi, a more recent presentation software, is intriguing. It is extremely dynamic, and more organic in its organizational structure. Used effectively, it can be very engaging to audiences. But unfortunately, I have seen too many Prezi presentations which were disorienting for the audience, and even the speaker. If you are going to design and develop a Prezi presentation, make sure that you follow effective design logic, so that both you and the audience know where you are going and why. Even more so that PowerPoint (which audiences are generally more familiar with), you must orient audiences to the structure of a Prezi presentation, especially by forecasting what is coming up next. Practice, practice, practice.

VIDEO

Video can be a versatile type of visual. But a problem is that too often speakers let the video *be* the speaker. The correct use of video for presentation purposes requires that you, again, prime the audience and that you talk about the video, explaining its relevance to the topic. If you use these visuals to your advantage, they can be very effective, stimulating questions and discussion.

USING VISUALS

Using visuals takes practice. While each type has advantages and disadvantages for presenting information, each also has its problems for the speaker in terms of using it "on the spot."

White boards are probably the easiest visual for speakers to use once the problem of writing on them comfortably is solved. No doubt this ease of use is why they are popular.

PowerPoint slides require that speakers double-check the order of the presentation. You do not want to have to apologize for the next slide not being the one you were expecting.

Video requires making sure that what you want to show is available and ready with only a click of a mouse. The last thing you want to subject an audience to is wandering through YouTube looking for something you saw while you were preparing the presentation and then not being able to find it. Locate it, save it, and have it ready . . . before you mention that you will show it as an example.

CONCLUSION

There are no perfect visuals. Each type has advantages and disadvantages, both in their ability to present information and in our ability to use them. In this chapter, we have examined different types of visuals available to us, looking at how they can add to our presentations and how they can detract from them if we are not comfortable with their use. That last point is the most important: Choose visuals that will add to your presentation, that will fit into the presentation environment you will be using, and that you feel comfortable using. Then—once again—practice, practice, practice!

CHAPTER 28

How to Run Effective Meetings

In all businesses, an important segment of work is carried on in meetings through small-group discussions. Often, those of us who have to sit through a lot of these sessions become convinced that there must be a more efficient way of running things. Too many meetings seem dedicated to the inverse law of meeting management: the longer the meeting, the less that gets done. Fortunately, it does not have to be this way. Successful meetings depend on the following factors:

- leadership ability
- problem-solving skills
- listening skills

Unfortunately, these skills are often thrown down and trampled on during most business meetings. This chapter will show you ways to run effective meetings. Figure 28.1 is an outline of how to do this.

Remember, however, that in small-group discussions, perhaps more than in any other type of communication, personality differences come into play. Different people react in different ways, and since most writing projects are the result of collaborative effort (interpreted as plenty of meetings), it would probably be a good idea to review the material in Chapter 1 to develop the interpersonal skills for handling these differences and for turning them into strengths.

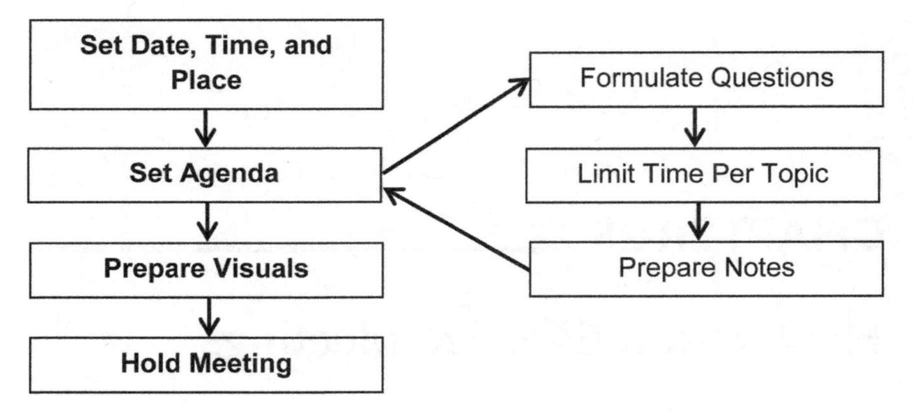

Figure 28.1. VOS for Meetings.

LEADERSHIP ABILITY

All groups have leaders. For planned meetings, the role of leader is generally assigned. In other words, if a manager calls a meeting, everyone knows who the leader is. With informal discussions, the situation is different and less rigid, but a leader will emerge.

Regardless of the meeting's purpose or environment, leaders do similar things. They set the tone of the meeting, making it serious or easygoing. They order the flow of business by manipulating the discussion. Good leaders use this control to solve group problems or to cover the topic to be discussed. Some leaders, however, do the same thing with only self-serving intentions, playing up their own sense of importance. For this reason, part of a meeting's success depends upon the leaders' integrity. No matter what the leaders' intentions might be, they order the flow of information in a meeting by interjecting questions and comments. In very formal meetings, this leadership is enhanced by a rigid agenda with time allotments for each topic to be considered.

Leaders also listen. They hear agreement among members of the group, aid exploration of what is agreed upon, and help make decisions.

PROBLEM-SOLVING SKILLS

Discussions that solve problems depend upon a willingness of all members of the group to communicate their ideas freely, to cooperate with others, to listen to their ideas, and to give and receive constructive criticism of ideas. In such an environment, group members contribute their knowledge to the problem-solving process. Every member of a group has something valuable to say, if for no other reason than to present a different viewpoint on a problem. Being able to criticize the ideas that are discussed and being able to receive criticism of your own ideas are the only ways to ensure this free interchange of information. Finally, cooperation among members of the group makes it likely that something useful will come out of the discussion.

To facilitate problem-solving within meetings, leaders should attempt the following:

- **Ask a question.** Asking a question as a way of organizing discussion is preferred over choosing a topic, because it focuses the group's attention on the problem. Questions have answers; topics only have subtopics. Choosing a topic or presenting a topic often leads to the seemingly endless, and almost always fruitless, discussions that all of us have experienced.
- **Define a question.** This requires the group to agree on precisely what is at issue. It builds a fence around what is to be discussed and enables leaders to keep the discussion centered on the problem.
- **Examine the background.** The only way problems can be solved is if the group understands them. Part of this understanding is knowing what led to the problem in the first place. This is an extension of defining the question.
- **Select criteria for evaluating suggested solutions.** This is another fence-building task. It enables the group to agree on what should be important in proposed solutions. Then if some solutions do not meet these criteria at the outset of the discussion, they can be discarded without hurt feelings.
- **Suggest solutions.** In successful problem-solving meetings, several alternatives will be examined.
- **Discard all but the best solution.** This is still the most time-consuming task of group problem-solving. Even in the best of situations, agreement will rarely be easy to come by. But following this procedure will at least ensure that the meetings you lead will arrive at this stage as efficiently as possible. The procedure enables groups to avoid wandering down unproductive tangents.

LISTENING SKILLS

Effective listening requires that meeting leaders let others talk. Limit responses and questions to the main point of discussion. Remember that you are not overbearingly imposing your will on the direction of the group, but are using your skills as a leader to keep the discussion focused.

Attempt to clarify what other members of the group have said. This summarizing allows time for the group to reflect on what has been discussed and to gather thoughts and evaluate. Resist telling the group what they should be thinking and saying. This restraint is hard for some types of people to achieve as leaders, but if you can master it, your leadership and problem-solving will be better as a result.

Try to understand solutions and ideas from the other group members' perspectives. This openness and willingness to discuss alternatives help involve all group members in the problem-solving process. This involvement is important, even if the final decision is yours to make.

Pay attention to courtesy. All of us have experienced situations in which we knew what the other person was going to say or how an idea was going to be developed. The temptation, particularly among people who are intuitives, is to interrupt and finish the other person's thought. Don't do it! Not only is this practice rude and disrespectful of the other person, it destroys the cooperation of the group. Do too much of this sort of thing, and other group members will quit speaking freely.

TECHNOLOGY-ENHANCED MEETINGS

Two technological meeting enhancements—conference calls and videoconferencing, using platforms such as Skype—create an entirely different meeting environment from having everyone in one physical space. In fact, we can think of it as the virtual meeting.

Conference calls, while they allow people in several locations to talk to each other at the same time, impose a serious limitation on meetings—the inability of any participant to experience nonverbal cues (facial expressions, body language) that are vitally important to face-to-face meetings. Conference calls are therefore usually best left for obtaining immediate answers to relatively short and simple questions, or for resolving very simple issues. Even though many organizations use them for a much broader range of meeting activities, including initial job interviews, we should recognize the significant omission of an important aspect of human communication—the ability to see each other.

Videoconferencing solves that problem. At its best, technologies such as Skype preserve the meeting environment by allowing participants to see each other and to benefit from visual cues. As such, they truly create a virtual environment for holding meetings, and as a result, all the advice for running effective meetings applies when using Skype or any other type of videoconferencing. But as is the case with a lot of technological advancement, reality is often less than the hype.

Despite the visual approximation of a face-to-face meeting, participants are still separated by many miles, time zones, and even political and cultural boundaries. These issues must be considered when holding a virtual meeting using videoconferencing. And too often, the technology can be a plague. Every one of us has had the experience of Skype seizing up—the image freezes, the audio goes out, people have chosen inappropriate backgrounds for "Skyping in." Just in my experience, these have included such telecommuting improprieties as dogs barking, teakettles whistling, cats jumping into laps, people dressed in all sorts of clothing they would never wear to the office, distracting visual backgrounds, and yes, even toilets flushing. In the end, while the use of Skype and other videoconferencing technologies is probably unavoidable in the 21st century, the primary reason for doing so is cost. It's cheap.

So, if videoconferencing is here to stay (until we better develop telepresence, haptic technology, etc.), at least use a modicum of judgment in how it is set up and used. Common sense should dictate that in videoconferencing we should be every bit as professional as if we were sitting in the same room.

Or, on the other hand, rather than e-mailing a colleague three cubicles down, you might get out of your chair, walk 40 feet, and meet face-to-face. The result could very well be more enjoyable and more productive.

CONCLUSION

This chapter has presented some of the techniques for leading effective meetings. It is not a complete discussion of this large task. Completeness would require a separate book, books that I have included in the suggested readings at the end of this chapter. Nonetheless, leading meetings and small-group discussions is a vitally important aspect of communicating within scientific and technological industries. Brainstorming design sessions are common. And more and more, much writing is done collaboratively. Problems are identified in meetings, solutions discussed, and documents reviewed. Too often, these meetings can be self-promoting affairs because of how strongly egos are wrapped up in the design or writing process. Subordinating one's ego to the group is not a denial of integrity. Rather, it is a realization that effective group thinking makes meetings more effective and shorter.

SUGGESTED READINGS

Kidd, Peter. *Powerfully Simple Meetings: Your Guide for Fewer, Faster, More Focused Meetings*. Meeting Result, 2014.

Running Meetings: Expert Solutions to Everyday Challenges. Boston: Harvard Business School Press, 2006.

3-M Meeting Management Team. *How to Run Better Business Meetings*. New York: McGraw-Hill, 1987.

PART IX

Conclusion

CHAPTER 29

Final Thoughts

In this book, chapter after chapter has described strategies for effective and successful communication, both written and spoken. These are powerful tools. They have stood the test of time, many dating back 2,400 years or so to their codification in Classical Greece by the pillars of world culture—Socrates, Plato, Aristotle. Undoubtedly, they were in practice long before that, most likely back to the emergence of civilization itself. And they work.

In an era where the new term "alternative facts" has arisen to describe untruth used in the pursuit of political gain, it is important to conclude a book such as this one with an examination of how to communicate ethically, using the strategies provided in the preceding pages. The reality is that these very same strategies, which are effective in professional environments to communicate about technology and science, can be (and still are) used more nefariously. They are the strategies used to fudge experimental results, to make overreaching promises about a new product, to pass blame for personal and professional failure onto others, to self-aggrandize when teamwork was responsible for success. News media is full of examples of persons using these strategies in precisely these ways, and in many cases, reporting on the results of getting caught—professional censure, demotion, even arrest. And at their darkest use throughout history, these strategies are also the strategies of propaganda.

ETHICS—A BRIEF SUMMARY

Ethics makes up an entire discipline of study; there are university courses and majors devoted entirely to its study. The earliest and still among the most important examinations of ethics was written by Aristotle over two millennia ago—*The Nichomachean Ethics*. As a result, it would be easy to devote an entire book to the topic, but that is not necessary here. Instead, we will examine some of the core ideas that can apply directly to communication about science and technology in professional environments. In doing so, I hope to avoid unfamiliar terminology as much as possible, but when it is unavoidable, I will explain what I mean and how it is applied.

One way to think about ethics is to see it in terms of the ever-present conflict between what can be done (legality) and what should be done (morality). While there are subcategories within the study of law often called "legal ethics," that term still refers to what can be done within the confines of existing law. But law, as history tells us, is permeable. For example, if we consider the long history of speech freedoms and the laws that regulate them, we can easily see that they have evolved and are still evolving.

Commercial speech is a broad category of communication that includes all professional uses (advertising, public relations, technical, scientific, professional, business communication, etc.). First Amendment law in the United States affords commercial speech fewer freedoms than political speech (upon which the authors of the First Amendment placed the highest value). As a result, in the United States, commercial speech can be regulated if it is misleading or concerns an illegal product. With the establishment of the Federal Trade Commission (FTC) in 1914, the U.S. government gave itself the power to regulate all forms of marketing communication, as well. Most often, this is applied to regulating unfair or deceptive acts or practices; in other words, marketing communication—including advertising—that is not true and, most importantly, potentially can harm consumers. In such a circumstance, the FTC can require an advertiser to stop making the deceptive claim about a product or service. Or if the commission considers the circumstance serious enough, it can force an advertiser to provide more (and truthful) information about the product or service. A good example of this is the warnings on cigarette packaging that began appearing in the 1960s. In the most serious of cases, the FTC can require a company to create advertising that corrects consumer beliefs about a product or service, based on deceptive advertising. Once again, using the tobacco industry, the "truth" advertising campaigns that began in the 1990s are an example of this.

But there is a form of stretching the truth that the FTC generally does not bother itself with. This is most often referred to as "puffery." Puffery is simple exaggeration or hyperbole—common in a lot of marketing communication—that is opinion-based, harmless persuasion and that most sensible consumers of that information recognize as such, rather than considering it to be deceptive. This sort of communication is not regulated. A good example to think of is any time you see or hear an advertisement that describes its product or service as "the best." My own personal favorite from my youth is the Ivory Soap claim that it is "99 and 44/100 percent pure." Pure what?

Or, even more seriously, consider the fact that virtually all cultures have legal proscriptions against killing, but that societies have developed a theory of just war to condone killing by a state when it is deemed necessary. These principles date back over 2,000 years but are most clearly explained in the writing of Thomas Aquinas in the 13th century.

What we think of when we use the term "ethics" primarily relates to morality, what *should* be done after a careful analysis of all potential actions. That is what we need to consider when using powerful communication strategies.

For much of the last century, most people who consider the topic of ethics are in relative agreement that the practice of moral values and prescriptive judgments ("thou shalt not . . .") is subjective to culture, time, and place. When one hears someone talking about relative morality or relative ethics, this is usually what is being referred to. And even though most people—who do not spend much time thinking about ethics—tend to react negatively to such a concept, it is in fact the reality of ethics. Like law, ethics has evolved through history. For example, "an eye for an eye"—an ethical principle of retribution for wrongs committed against an individual—was common across cultures early in the development of civilization, but is no longer viewed as an ethical component of modern societies.

Perhaps a less judgmental way to think about relative ethics is the following example: Most of us might agree with the statement "Hunger must be abolished." But abolishing hunger leads to greater survival rates, and greater survival rates lead to a growth in population, and increased population leads to a reduction in available resources, and the reduction in available resources leads to an increase in hunger.

So, what is a professional who communicates regularly about science and technology to do?

Developing Professional Ethical Standards

A good way to begin thinking about your own professional ethical standards is to consider what is often called the ethical circles of influence.

The centermost circle reflects your own personal and professional standards of behavior. At the second level outward are the standards of the organization for which you work. At the third level are the ethical standards of your profession (almost all professions have these, and they can easily be researched). At

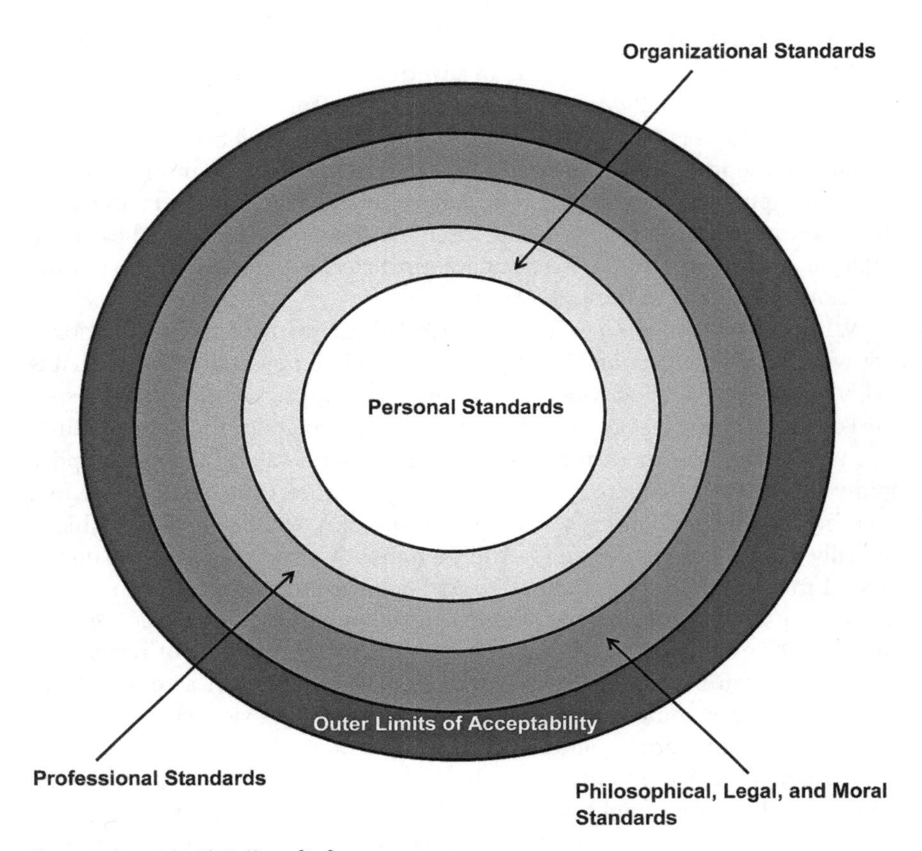

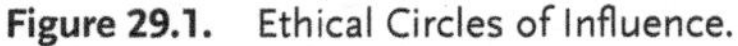

Figure 29.1. Ethical Circles of Influence.

the fourth level are disciplines such as philosophy, rules, and laws (many of which were summarized in the earlier part of this chapter, as they apply to communicating about science and technology). And at the final level, one finds the outer limits of what readers, consumers, the public, will tolerate. To accurately use a common pop-culture phrase, "you don't want to go there" regarding ethical professional communication.

Another way to approach developing your own professional standards of communication behavior is to apply a long-valued, four-strategy approach to how you should respond to expectations—communicative or otherwise—that you feel violate your ethical standards.

Let's say that you feel compelled (implicit request) or you are asked (explicit request) to do or communicate something that violates your professional standards of ethical behavior.

The first and highest-valued strategy is to educate your organization to accepting your standards of ethical behavior. The thought behind this is that doing so will improve the values of the organization, its perceived reputation by the public, and its behaviors. To do this, you must have a soundly reasoned argument, applying the persuasive principles described elsewhere in this book. You must realize that the attempt to apply this strategy often means confronting the highest level of management in your organization, as well as challenging long-standing organizational culture. But this strategy also adheres to the highest level of professional ethical behavior.

The second-most-valued strategy is to explain that the assignment is counter to your professional standards of behavior and ask for an alternative assignment. Applying this strategy again requires you to be very persuasive in explaining why it should be followed. But it also pinpoints your own professional value to your organization. And the result of applying this strategy—whether it is successful or not—certainly clarifies for you the values of the organization you work for.

The third strategy is to refuse to do what you believe to be unethical. Applying this strategy can lead to your dismissal from the organization. But perhaps this is a blessing in disguise, since the entire affair might demonstrate a poor fit between you and your employer.

And the fourth strategy is to do it, regardless of your ethical principles. If there is a positive side to this, it is that you demonstrate to your organization that you are a team player who will not rock the boat, but it does so by invalidating your ethical standards.

Obviously, each of these strategies contains risk. The first three might result in your losing your job; the fourth strategy results in you losing your ethical principles. You might think that—ideally—the choices are clear to you. But consider the reality of the following: You are confronted with the possibility of doing or communicating something that you feel may violate your professional ethical standards. You have just bought a house and incurred a 30-year mortgage. Your first child is on the way. Now, what do you do? Ethics *are* relative, and they are never easy.

To finish this chapter, and this book, I have included the Code of Ethics from the Society for Technical Communication (STC). It is an excellent standard for anyone who engages in communicating about science and technology. It can work for your organization, or it can be the starting point for developing a Code of Ethics that applies specifically to your organization. It also is a good summary of the principles we have been discussing in this chapter.

As Technical Communicators, We Observe the Following Ethical Principles in Our Professional Activities.

Legality

We observe the laws and regulations governing our profession. We meet the terms of contracts we undertake. We ensure that all terms are consistent with laws and regulations locally and globally, as applicable, and with STC ethical principles.

Honesty

We seek to promote the public good in our activities. To the best of our ability, we provide truthful and accurate communications. We also dedicate ourselves to conciseness, clarity, coherence, and creativity, striving to meet the needs of those who use our products and services. We alert our clients and employers when we believe that material is ambiguous. Before using another person's work, we obtain permission. We attribute authorship of material and ideas only to those who make an original and substantive contribution. We do not perform work outside our job scope during hours compensated by clients or employers, except with their permission; nor do we use their facilities, equipment, or supplies without their approval. When we advertise our services, we do so truthfully.

Confidentiality

We respect the confidentiality of our clients, employers, and professional organizations. We disclose business-sensitive information only with their consent or when legally required to do so. We obtain releases from clients and employers before including any business-sensitive materials in our portfolios or commercial demonstrations or before using such materials for another client or employer.

Quality

We endeavor to produce excellence in our communication products. We negotiate realistic agreements with clients and employers on schedules, budgets, and deliverables during project planning. Then we strive to fulfill our obligations in a timely, responsible manner.

Fairness

We respect cultural variety and other aspects of diversity in our clients, employers, development teams, and audiences. We serve the business interests of our clients and employers as long as they are consistent with the public good. Whenever possible, we avoid conflicts of interest in fulfilling our professional responsibilities and activities. If we discern a conflict of interest, we disclose it to those concerned and obtain their approval before proceeding.

Professionalism

We evaluate communication products and services constructively and tactfully, and seek definitive assessments of our own professional performance. We advance technical communication through our integrity and excellence in performing each task we undertake. Additionally, we assist other persons in our profession through mentoring, networking, and instruction. We also pursue professional self-improvement, especially through courses and conferences.

Adopted by the STC Board of Directors

September 1998

Figure 29.2. Ethical Principles, Society for Technical Communication. (Society for Technical Communication. Ethical Principles. https://www.stc.org/about-stc/ethical-principles/. Used by permission.)

Index

About the Author

Charles H. Sides, PhD, holds the rank of Professor and directs the internship program for the Department of Communications Media at Fitchburg State University. A recognized scholar in applied rhetoric, he has published 10 books and numerous articles on communication issues, including *The Right to Write: College Communication and the First Amendment, Freedom of Information in a Post 9–11 World* and *Internships: Theory and Practice.* He is Executive Editor of the *Journal of Technical Writing and Communication (JTWC)* and Co-Editor of The Routledge Series on Technical Communication, Rhetoric, and Culture. As Editor of Baywood's Technical Communication Series for over 20 years, he managed to publication 42 books, several receiving national awards of excellence. A consultant to defense, high-tech, medical, and publications industries, he has worked with clients across the United States, the Middle East, and the Far East.